GARDENER TO GARDENER™

GARDENER TO GARDENER™

Seed-Starting Primer & Almanac

Hundreds of Great Ideas, Tips, and Techniques from Gardeners Just Like You

Edited by Vicki Mattern

RODALE®

WE INSPIRE AND ENABLE PEOPLE TO IMPROVE
THEIR LIVES AND THE WORLD AROUND THEM

We're always happy to hear from you. For questions or comments concerning the editorial content of this book, please write to

Rodale Book Readers' Service
33 East Minor Street
Emmaus, PA 18098

Look for other Rodale books wherever books are sold. Or call us at (800) 848-4735.

For more information about Rodale Organic Living magazines and books, visit us at

www.organicgardening.com

Editor: Vicki Mattern
Project Manager: Karen Bolesta
Cover and Interior Book Designer: Nancy S. Biltcliff
Contributing Designer: Gavin Robinson
Cover and Interior Illustrator: Keith Ward
Layout Designer: Daniel MacBride
Researcher: Diana Erney
Copy Editors: Roger Yepsen and Jennifer Blackwell
Editorial Intern: Megan O'Connell
Product Specialist: Jodi Schaffer
Indexer: Nanette Bendyna
Editorial Assistance: Susan L. Nickol and Dolly Donchez

RODALE ORGANIC LIVING BOOKS
Editorial Director: Christopher Hirsheimer
Executive Creative Director: Christin Gangi
Executive Editor: Kathleen DeVanna Fish
Art Director: Patricia Field
Content Assembly Manager: Robert V. Anderson Jr.
Studio Manager: Leslie M. Keefe
Copy Manager: Nancy N. Bailey
Projects Coordinator: Kerrie A. Cadden

Library of Congress Cataloging-in-Publication Data

Gardener to gardener seed-starting primer & almanac : hundreds of great ideas, tips, and techniques from gardeners just like you / edited by Vicki Mattern.
          p.   cm.
     Includes bibliographical references (p.   ).
     ISBN 0–87596–872–4 (hardcover : alk. paper)
     ISBN 0–87596–884–8 (pbk. : alk. paper)
     1. Organic gardening.   2. Seeds.   I. Title: Seed-starting primer & almanac.   II. Mattern, Vicki.
     SB453.5 .G34 2001
     635'.0484—dc21   2001004673

Distributed in the book trade by St. Martin's Press

2 4 6 8 10 9 7 5 3 1 hardcover

2 4 6 8 10 9 7 5 3 1 paperback

RODALE
Organic Gardening Starts Here!

Here at Rodale, we've been gardening organically for more than 60 years—ever since my grandfather J. I. Rodale learned about composting and decided that healthy living starts with healthy soil. In 1940 J. I. started the Rodale Organic Farm to test his theories, and today the nonprofit Rodale Institute Experimental Farm is still at the forefront of organic gardening and farming research. In 1942 J. I. founded *Organic Gardening* magazine to share his discoveries with gardeners everywhere. His son, my father, Robert Rodale, headed *Organic Gardening* until 1990, and today a third generation of Rodales is growing up with *OG* magazine. Over the years we've shown millions of readers how to grow bountiful crops and beautiful flowers using nature's own techniques.

In this book, you'll find the latest organic methods and the best gardening advice. We know—because all our authors and editors are passionate about gardening! We feel strongly that our gardens should be safe for our children, pets, and the birds and butterflies that add beauty and delight to our lives and landscapes. Our gardens should provide us with fresh, flavorful vegetables, delightful herbs, and gorgeous flowers. And they should be a pleasure to work in as well as to view.

Sharing the secrets of safe, successful gardening is why we publish books. So come visit us at www.organicgardening.com, where you can tour the world of organic gardening all day, every day. And use this book to create your best garden ever.

Happy gardening!

Maria Rodale

Maria Rodale
Rodale Organic Gardening Books

Contents

Welcome to
Gardener to Gardener!

Once you discover the joys of gardening, you'll realize that gardening can be a fulfilling year-round activity. You'll start to fill your calendar each month with to-do lists, and look beyond the growing season months of June, July, and August to find tasks that will keep your green thumb green, even when the cold winds blow. If you're awe-inspired by the beauty and power of Nature when you see the first tender shoots emerge from the soil in spring, you definitely know you've been bitten by the gardening bug. When you harvest a basket of chili peppers and spend a moment admiring the pile of brilliant colors before you get to work making pepper-onion relish, rest assured that there will be more ripe peppers in just a day or so. And if you wonder how you can fit in all your gardening plans before daylight wanes, you know that gardening is becoming a passion. Don't worry; you're not alone. Millions of folks garden, and they have always relied on each other for advice about starting seeds, controlling pests, and improving the soil.

Let an Almanac Be Your Guide

Besides exchanging tips and secrets over the garden gate, gardeners have been keeping records about their gardens in journals and almanacs for generations, and each discovery helps a whole new generation of gardeners. That's why we think you'll find this *Gardener to Gardener* almanac especially useful. It's organized like a calendar, just like a traditional almanac, and includes a "Gardener's To-Do List" for each month. Because your local climate affects when and how you garden, we've organized these gardening checklists according to the USDA Plant Hardiness Zones. (If you don't know what zone your garden is, check the hardiness map on page 230 to find out.) You'll also find other great activities you can do in your garden for each month. Plus, some chapters include log sheets you can photocopy to keep track of your seed inventory or to remind you of the dates you planted seeds.

This book features an in-depth look at seed starting. In Chapter 2, we take you step by step through the process of sowing seeds indoors and raising healthy seedlings organically. You'll find a discussion on seed-starting supplies, a guide to special seed-starting techniques, and a problem-solving section. There's even a plant-by-plant guide with hints for starting everything from ageratums to zinnias.

If you're a Deep South gardener (and live in Zone 10 especially), your gardening activities may be on quite a different schedule. When other gardens go dormant in the fall, yours is just coming to life, so if you garden in Zone 10, you'll need to apply some of the seasonal advice according to your own timetable.

A Trusted Source

Reading *Gardener to Gardener* is like talking to the most experienced gardener you know. The book is full of tips and hints from dozens of gardening friends: the readers of *Organic Gardening* magazine. This group of experienced organic gardeners has shared their best ideas on everything from making seed tape for tiny seeds to growing tomatoes year-round.

If you enjoy the reader tips in this book, you may also want to visit a special place online where organic gardeners meet, swap secrets, and talk about issues that affect their gardens. It's the Gardener to Gardener Forum at www.organicgardening.com. Check it out online!

Chapter 1

January

Getting Started: Planning and Designing the Garden

There are two seasonal diversions that can ease the bite of any winter. One is the January thaw. The other is the seed catalogues.

—Hal Borland

What do you need to create your dream garden? Nothing more than a pencil, some paper, and a few garden catalogs! With these simple tools and your own imagination, you can build beds, move shrubs, install a water garden, create a habitat for wildlife, or add any other garden feature you like. Of course, it's only a plan. But planning is the first—and arguably most important—step toward making your dream garden a reality.

Throughout most regions, January is the perfect time for garden planning. Landscapes are easier to assess and gardeners have more time to assess them. And the new crop of garden catalogs provides plenty of inspiration and information to get the process going.

After you've decided on the broad outline of your garden for this year, you can zero in on specific plant choices, taking into consideration last year's successes and failures. Did your tomatoes succumb to disease before you could pick your share of ripe fruits? Then be sure to choose a tomato variety that offers some disease resistance. Did your canna lilies fail because the site was too shady? Maybe this year you should try astilbes or another shade-loving plant (or resolve to thin that overgrown tree!). Once you've looked back, you can look forward to a whole new year of garden fun!

Gardener's To-Do List—January

**If you don't know what USDA hardiness zone you live in,
check the map on page 230 to find out.**

Zone 3

- [ ] For an early taste of spring, grow some sprouts. Mung bean, radish, and buckwheat sprouts grow well in vented jars. Just put the seeds inside, cover them with water overnight, drain, then rinse twice a day.
- [ ] Explore seed catalogs, then send in your orders.
- [ ] Start snapdragon and pansy seedlings.

Zone 4

- [ ] Check the viability of old seeds by sprouting a few of each kind in damp paper towels enclosed in plastic bags.
- [ ] Set up your seed-starting system.
- [ ] Start a flat of hardy perennials or alpine strawberries.
- [ ] Rearrange houseplants so that all get their share of bright light. Later in the month, give them a light feeding.

Zone 5

- [ ] Gather your seed-starting equipment, then start seeds of pansies, snapdragons, and hardy perennials.
- [ ] Toward the end of the month, start onion seeds.

- [ ] Order seeds of cabbage, broccoli, cauliflower, parsley, and peas.
- [ ] Dig up a frozen chunk of chives and force them into early growth indoors.
- [ ] When the snow cover is thin, check shrubs and perennials. If the roots have heaved up too close to the surface, press them back into place.
- [ ] Late this month, begin pruning apples.
- [ ] If rabbits are ravaging fruit trees and shrubs, lure them to a distant spot with corn or hay.

Zone 6

- [ ] Start hardy flowers under lights.
- [ ] In midmonth, start cabbage and onion seeds indoors.
- [ ] Order seeds for other cool-weather crops, such as broccoli, cauliflower, spinach, celery, lettuce, and peas.
- [ ] Get your coldframe ready. Mound a 4-inch layer of leaves or soil around the outside to help it hold heat.
- [ ] Check winter mulches and replenish those that have thinned.
- [ ] Wrap wire mesh around tree trunks that have been damaged by rodents.

Zone 7

- [ ] Start seeds of cabbage, onions, and hardy herbs under bright lights early this month.
- [ ] Clean out your coldframe.
- [ ] Collect plastic jugs to use as cloches.
- [ ] Late this month, mow winter cover crops.
- [ ] Direct-seed sweet peas.
- [ ] Indoors, start seeds of perennials, such as columbine and balloon flower.
- [ ] Begin dividing daylilies.
- [ ] Prune crape myrtles.
- [ ] Set out junipers, hollies, and other evergreens.

Zone 8

- [ ] Harden-off cool-weather transplants (cabbage, broccoli, etc.) that you've started indoors.
- [ ] Start lettuce indoors. When seedlings have several leaves, set them out under plastic milk jug cloches.
- [ ] Sow peas outdoors late this month.
- [ ] Set seed potatoes in a bright spot to encourage sprouting.
- [ ] Take advantage of your last chance to dig and divide crowded daylilies and daffodils.
- [ ] Dig and transplant dormant phlox, thrift, and hosta.
- [ ] Trim old leaves from liriope, but don't disturb the crowns.

Zone 9

- [ ] Sow beets, carrots, lettuce, peas, and spinach in the garden.
- [ ] Indoors, start seeds of tomatoes, peppers, and eggplants.
- [ ] Direct-seed alyssum, California poppies, nasturtium, and cornflowers.
- [ ] Prune geraniums to stimulate bushy new growth.
- [ ] Plant new persimmons, loquats, and figs.
- [ ] In the West, plant young pistachios in well-drained soil, then stake them securely.

Zone 10

- [ ] Set out transplants of onions, potatoes, cabbage, and broccoli.
- [ ] Water transplants often to keep them growing strong.
- [ ] Plant dahlias, caladiums, gladiolus, and tropical tubers, such as amaryllis and crinums.
- [ ] Set out bedding plants such as pansies, dianthus, and petunias.
- [ ] Plant young plumerias in containers.
- [ ] Trim back poinsettias and other tropicals after they finish blooming.
- [ ] Taste citrus fruits regularly so that you can harvest them at their peak of flavor.

Location Matters
If your landscape is hilly, don't put your garden at the top of a slope, where it's likely to be windy and cold. Also avoid the bottom of the slope, where water and cold air collect. Instead, position your garden on a warm, well-drained site on the face of a slope.

Getting Started

Gardening organically is really quite simple. Push a seed into the soil, give it sun and water, add compost, and keep down the weeds. The seed knows what to do. It will sprout, soak up nutrients, develop a stem and leaves, and before you know it, reward you with flowers, fruits, and seeds.

Your role is to nurture it along using techniques adapted straight from nature. There's nothing difficult or artificial; there are no expensive chemicals to buy. Follow this simple advice and experience the joy of growing food and flowers the healthful way—without toxic pesticides and fertilizers.

Site It Right

When choosing a new garden site or expanding an existing one, take a close look at the surrounding landscape—especially at any nearby trees. Estimate how big those trees will be in 5 to 10 years, then plan your garden so that it won't be shaded when the trees get larger.

If you are able to garden year-round, be sure to consider where the sunlight will hit your garden in winter, too. If possible, site your garden on the south side of a building to take full advantage of the winter sun.

No matter where you live, try to orient vegetable garden beds to run from north to south. That way, taller crops won't shade shorter ones however you plant them.

Great gardens start with solid planning. From choosing time-tested varieties and starting seeds indoors to amending the soil with compost and cultivating before planting, your recipe for success is equal parts indoor homework and outdoor planting.

Raise Your Beds

The loose soil of a well-prepared raised bed allows plant roots to spread out and take in the maximum amount of nutrients and water. Raised beds also warm up faster in spring and drain better than flat beds. And because you don't walk on them (you work them from the sides), the soil never becomes compacted.

To keep plants within easy reach, make the beds no wider than *double your arm length* (about 4 to 5 feet). The length of the beds depends on the size of your garden and the lay of your land. Shorter beds can save time when weeding or harvesting, but the paths around lots of short beds can take up valuable garden space if you have a small garden. Many gardeners find 15- to 20-foot-long beds to be about the right size for convenience and efficient use of space.

If you have a structure (such as a fence or building) on one side of your garden, put a narrow bed along it rather than a path. Maneuvering a wheelbarrow down a path next to a fence or building can be difficult.

There are many methods for making raised beds. For one of them, see "A Simple Way to Raise Your Beds" on page 72.

Plan Perfect Paths

When creating your garden plan, consider the width of paths you'll need to accommodate your equipment. If you plan to use a wheelbarrow, tiller, or cart, you'll need at least a few main paths that are 18 to 24 inches wide. Smaller, secondary paths can be narrower—12 to 18 inches wide.

To keep the weeds out of those paths, mulch them with a thick layer of clean straw, hay, or wood chips. Or simply make them wide enough for a lawn mower and grow grass on them. Mow the paths regularly and use the clippings to mulch the beds.

Keep the Compost Close

Compost is a must for any organic garden! It contains all of the primary nutrients, trace minerals, and beneficial organisms that plants need to grow and thrive. It also discourages plant diseases and neutralizes the chemistry of the soil. If your soil is too acidic or too alkaline, you can help put it back into balance by adding compost. (For tips on making great compost, see Chapter 3.)

Plan to locate your compost pile or bins as close as possible to the garden. You'll be more inclined to use the compost, and you'll save time and effort because you won't need to do as much hauling. And when summers are dry, you can water the compost at the same time you water your garden.

For New Gardeners

If this is your first garden, stick with easy, surefire plants such as tomatoes, leaf lettuce, zucchini, marigolds, and impatiens. Talk to experienced gardeners in your area for their suggestions on the best varieties for your conditions.

If you live in the South, look for varieties that are able to withstand the heat. Or, if you garden in the North and have a short growing season, choose varieties that will withstand the cold and mature quickly. Many seed catalogs indicate adaptability to climate in their variety descriptions.

And don't be discouraged if you have a few losses. If you continue to care for your soil and plants organically, your garden will become healthier, more productive, and more beautiful with each passing year.

Top 10 Tips for Seed Shopping by Mail

Seed shopping by mail can certainly liven up a dreary winter's day. These guidelines will help you make the best seed choices for your garden.

1. Peruse a variety of seed catalogs to compare selection and prices.

2. Start seed shopping in winter so you can work out the details of your seed-sowing and transplanting schedule.

3. Keep all seed catalogs for use as reference once you begin planting.

4. Look for catalogs that specialize in plants that grow well in your specific region of the country.

5. Weigh the amount of variety you want versus the price of the seed packets. Small seed packets of individual plants are reasonably priced but only contain one type of plant. Seed mixtures give you many types of plants in a single packet but usually cost significantly more.

6. Realize that the number of days to maturity quoted in catalogs is an estimate. The actual number of days may be slightly different for your area.

7. Decide between hybrids and open-pollinated cultivars. Hybrids may produce earlier harvests and higher yields, but open-pollinated cultivars may taste better, produce over a longer season, and usually cost less.

8. Watch out for seeds that are treated with synthetic chemical fungicide. When ordering, specify untreated seeds.

9. Choose cultivars that have qualities that are important to you, such as plant size, habit, and tolerance of your soil conditions.

10. Look for All-America Selections because these tend to grow and produce well in a variety of conditions.

Seed-Starting Glossary

These explanations will help you understand some of the words and terminology used in seed and plant catalogs.

Amendment \ə-'men(d)-mənt\:
Material that improves soil condition and aids plant growth.

Bolt \'bōlt\:
To produce flowers and seed prematurely, usually due to hot weather.

Cotyledon \ˌkä-tə-'le-dᵊn\:
Also called seed leaves. The leaf (or leaves) present in the dormant seed that unfolds as a seed germinates. Cotyledons often look different than the leaves that follow them. In seeds such as beans, they contain stored nutrients.

Damping-off \ˌdam-piŋ-ȯf\:
A disease caused by various fungi that results in seedling stems that shrivel and collapse at soil level.

Direct-seed \də-'rekt-'sēd\:
To sow seeds outdoors in garden soil.

Dormancy \'dȯr-mən(t)-sē\:
A state of reduced biochemical activity that persists until certain conditions occur that trigger germination.

Hardening off \'här-dᵊn-iŋ-'ȯf\:
Allowing young plants to adapt slowly to wind, temperature, and light conditions outside by increasing the amount of time the plants spend outdoor each day.

Scarification \ˌskar-ə-fə-'kā-shən\:
Nicking or wearing down hard seed coats to encourage germination.

Seed \'sēd\:
A plant embryo and its supply of nutrients, often surrounded by a protective seed coat.

Seed germination \'sēd ˌjər-mə-'nā-shən\:
The beginning of growth of a seed.

Seedling \'sēd-liŋ\:
A young plant grown from seed. Commonly, plants grown from seeds are termed seedlings until they are first transplanted.

Stratification \ˌstra-tə-fə-'kā-shən\:
Exposing seeds to a cool (35° to 40°F), moist period to break dormancy.

Viable \'vī-ə-bəl\:
Capable of germinating.

7 Secrets for a High-Yield Vegetable Garden

Imagine harvesting nearly half a ton of tasty, beautiful, organically grown vegetables from a 15 × 20-foot plot, 100 pounds of tomatoes from just 100 square feet (a 4 × 25-foot bed), or 20 pounds of carrots from just 24 square feet.

Yields like these are easier to achieve than you may think. The secret to superproductive gardening is taking the time *now* to plan strategies that will work for your garden. Here are seven high-yield strategies gleaned from gardeners who have learned to make the most of their garden space.

1. Build up your soil. Expert gardeners agree that building up the soil is the single most important factor in pumping up yields. A deep, organically rich soil encourages the growth of healthy, extensive roots that are able to reach more nutrients and water. The result: extralush, extra-productive growth aboveground.

The fastest way to get that deep layer of fertile soil is to make raised beds (as explained in Chapter 4). Raised beds yield up to four times more than the same amount of space planted in rows. That's due not only to their loose, fertile soil but also to efficient spacing—by using less space for paths, you have more room to grow plants.

Raised beds save you time, too. One researcher tracked the time it took to plant and maintain a 30 × 30-foot garden planted in beds, and

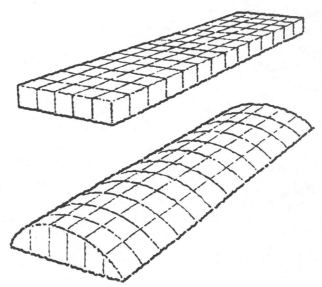

Rounding the surface of a 5-foot-wide flat bed (*top left*) increases the width of your planting surface to 6 feet (top right). Multiply that by the length of your bed for the total amount of space gained.

found that he needed to spend just 27 hours in the garden from mid-May to mid-October. Yet he was able to harvest 1,900 pounds of fresh vegetables—that's a year's supply of food for three people from about three total days of work!

How do raised beds save so much time? Plants grow close enough together to shade out competing weeds, so you spend less time weeding. The close spacing also makes watering and harvesting more efficient.

2. *Round out your beds.*

The shape of your beds can make a difference, too. Raised beds are more space efficient if the tops are gently rounded to form an arc, rather than flat. A rounded bed that is 5 feet wide across its base, for instance, will give you a 6-foot-wide arc above it—creating a planting surface that's a foot wider than that of a flat bed. That foot might not seem like much, but multiply it by the length of your bed and you'll see that it can make a big difference in total planting area.

In a 20-foot-long bed, for example, rounding the top increases your total planting area from 100 to 120 square feet. That's a 20 percent gain in planting space in a bed that takes up the same amount of ground space! Lettuce, spinach, and other greens are perfect crops for planting on the edges of a rounded bed.

To take advantage of that extra space gained when rounding a bed, plant shallow-rooted plants, such as lettuce greens and small herbs, along the bed's edges.

No Room for a Vegetable Garden?

You can plant a vegetable garden even if you have a small yard. The most important point to remember is to find the sunniest spot you can. If you don't have one large area that receives full sun, be creative and plant a few vegetables in a sunny spot along your driveway or mix them into your flowerbeds. (Many vegetable cultivars have beautiful leaves and flowers.) Growing vegetables in containers on your deck or roof is an option, too. Good choices for containers include lettuces, hot peppers, dwarf (or compact) tomatoes, and herbs.

Succession Planting, Southern Style

Here in Jacksonville, the garden season begins around January 1, when I plant cold-weather crops, such as radishes, beets, carrots, lettuce, and spinach. To prepare my raised beds for planting, I turn the soil, level it, and cover it with an inch of compost. Then I scrape back the compost where I want to plant, sow the seed, and cover the planted rows with a fine layer of compost. The plants germinate and grow fast. By the time they're harvested, pepper and tomato transplants, beans, and other warm-weather crops are ready to be planted. Best of all, with all of that wonderful compost, I need no additional fertilizer—and I'm munching fresh radishes by Groundhog Day!

Edward Luttrell
Jacksonville, Florida

3. *Space smartly.* To get the maximum yields from each bed, pay attention to how you arrange your plants. Avoid planting in square patterns or rows. Instead, stagger the plants by planting in triangles (as shown here). By doing so, you can fit 10 to 14 percent more plants in each bed.

Just be careful not to space your plants *too* tightly. Some plants won't reach their full size—or yield—when crowded. For instance, when one researcher increased the spacing between romaine lettuces from 8 to 10 inches, the harvest weight per plant doubled. (Remember that *weight* yield per square foot is more important than the number of plants per square foot.)

Overly tight spacing can also stress plants, making them more susceptible to diseases and insect attack.

Triangular spacing allows you to grow more in the same area.

4. *Grow up!* No matter how small your garden, you can grow more by going vertical. Grow space-hungry vining crops—such as tomatoes, pole beans, peas, squash, melons, cukes, and so on—straight up, supported by trellises, fences, cages, or stakes.

Growing vegetables vertically also saves time. Harvest and maintenance go faster because you can see exactly where the fruits are. And upward-bound plants are less likely to be hit by fungal diseases thanks to the improved air circulation around the foliage.

Try growing vining crops on trellises along one side of raised beds, using sturdy end posts with nylon mesh netting or string in between to provide a climbing surface. Tie the growing vines to the trellis. But don't worry about securing heavy fruits—even squash and melons will develop thicker stems for support.

5. *Mix it up.* Interplanting compatible crops saves space, too. Consider the classic Native American combination, the "three sisters"—corn, beans, and squash. Sturdy cornstalks support the pole beans, while squash grows freely on the ground below, shading out competing weeds. This combination works because the crops are compatible. Other compatible combinations include tomatoes, basil, and onions; leaf lettuce and peas or brassicas; carrots, onions, and radishes; and beets and celery.

6. *Succeed with successions.* Succession (or relay) planting allows you to grow more than one crop in a given space over the course of a growing season. That way, many gardeners are able to harvest three or even four crops from a single area.

For instance, an early crop of leaf lettuce can be followed with a fast-maturing corn, and the corn followed by more greens or overwintered garlic—all within a single growing season.

To get the most from your succession plantings:

- Use transplants. A transplant is already a month or so old when you plant it, and so will mature that much faster than a direct-seeded plant (one grown from seeds sown in the garden).

- Choose fast-maturing varieties.

- Replenish the soil with a ¼- to ½-inch layer of compost (about 2 cubic feet per 100 square feet) each time you replant. Work it into the top few inches of soil.

7. *Stretch your season.* Adding a few weeks to each end of the growing season can buy you enough time to grow yet another succession crop—say a planting of leaf lettuce, kale, or turnips—or to harvest more end-of-the-season tomatoes.

To get those extra weeks of production, you need to keep the air around your plants warm, even when the weather is cold, by using mulches, cloches, row covers, or coldframes.

Or give heat-loving crops (such as melons, peppers, and eggplants) an extra-early start by using two "blankets"—one to warm the air and one to warm the soil in early spring. About 6 to 8 weeks before the last frost date, preheat cold soil by covering it with either infrared-transmitting (IRT) mulch or black plastic, which will absorb heat. Then, cover the bed with a slitted, clear plastic tunnel. When the soil temperature reaches 65° to 70° F, set out plants and cover the black plastic mulch with straw to keep it from trapping too much heat. Remove the clear plastic tunnel when the air temperature warms and all danger of frost has passed. Install it again at the end of the season, when temperatures cool.

Top Crops for Small Spaces

If you don't plan to use a trellis but still want to grow vining crops, such as pumpkins, melons, or squash, consider cultivars with short vines. The plants take up less space in the garden, but their fruits are as large as those on standard cultivars. Examples include:

- 'Salad Bush' or 'Little Leaf' cucumbers

- 'Spirit' or 'Wizard' pumpkins

- 'Delicata' winter squash

- 'Minnesota Midget' cantaloupe

But keep in mind that you won't gain much in terms of yield with compact varieties of other crops. Bush beans, for example, don't produce as much per square foot as trellised pole beans. And miniature carrots take up less space, not only in the garden but also in the harvest basket.

Choosing Bloom Colors

Bloom colors cover the entire palette, from quiet pastels to bold, vibrant hues. When choosing colors for your garden, keep in mind that warm colors—reds, oranges, and yellows—come forward or jump out at you. Cool colors—greens, blues, and purples—retreat from the eye and have a more calming effect. If your garden is far from your viewing point, choose bright colors or create more contrast. If you'll look at your garden at close range, go for a more subtle color scheme. Create the illusion of depth in a small or shallow garden by using blue and purple flowers toward the back. For excitement, use bright colors or add a dash of red to a more subdued combination.

Designing for Beauty

Many gardeners know a beautiful garden when they see one, yet don't know the design elements that contribute to that look. Well, here's help—10 basic design tips to help you create a more beautiful garden.

*1. **Have a plan.*** Whether you're starting from scratch or rehabilitating an existing landscape, resist the temptation to buy plants on impulse, then drop them into the garden wherever you can find an opening. Instead, decide what you want your garden to be before you buy the plants. Sketch your plan on paper, then update it each season. One way to do this is with a "bubble diagram," shown here.

*2. **Start with the bones.*** First, decide where you want trees and shrubs to be. Then work in large-foliage plants, such as hostas. Last, add flowering perennials and annuals to the design.

*3. **Think in three dimensions.*** Lay out your garden beds so that they have a foreground, middle, and background. This provides depth and scale, and makes space seem larger than it is.

*4. **Make plantings big, bold, and simple.*** When creating your design, remember that large clusters of a single type of plant give a more pleasing effect than many different types of plants in the same amount of space.

*5. **Don't forget to be a little odd.*** Arrange plants and rocks in groups of three, five, or seven. An odd number of items tends to look more interesting than an even number.

To make a bubble diagram, begin with an idea for the size and shape of your garden, then draw its outline to scale on graph paper (1 inch on paper to 1 foot of ground is a good scale for most gardens). Use a pencil so you can move things around easily, and don't be afraid to make changes. Then, draw in circles or "bubbles" to represent your plants, making the scale of each bubble match the mature width of the plant. Write the plant name, color, and bloom season in each bubble if you have room, so you can see at a glance how good the combinations are likely to look.

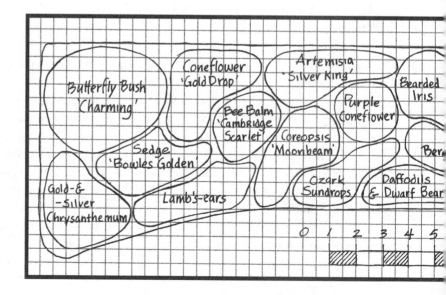

6. *Spread it out.* One of the most common design mistakes is planting too tightly. By the second season, perennials are hopelessly crowded. Instead, try to think 5 years ahead when leaving space for perennials and shrubs. To fill in the spaces and add drama in the meantime, use bold annuals such as sunflowers or nasturtiums.

7. *Tie it all together.* Pull together the disparate elements of your garden with a continuous, flowing form, such as a hedge, path, fence, or stone wall. Also, duplicate key plantings throughout the garden.

8. *Cool it with the colors.* Bright colors can be fun and exciting in the garden, but don't go overboard—too much color can result in a garish overstatement. Think fewer flowers and more foliage, especially foliage that has unusual textures and shapes.

9. *Be true to your locale.* Choose plants that are suited to your conditions. Consider your hardiness zone, annual rainfall, soil, and sunlight. If you've lived in your house only a few months, don't be in a rush to plant your garden—wait until you have a better handle on the site's conditions, as well as its "personality." Explore the natural areas around your home to learn what types of plants are best adapted to your locale—these "natives" are also likely to thrive in your home garden.

10. *Go with the flow.* Instead of trying to disguise or eliminate natural landscape elements, such as a rock, stump, or slope, use them to your advantage. Natural features add drama to an otherwise typical landscape.

Plants with Fabulous Foliage

For multiseason interest in your landscape, be sure to include plants with interesting foliage color or texture. Here are just a few of the many choices available:

***Acanthus* spp. (bear's breeches):** Shiny; lobed or heart-shaped; spiny

***Ajuga reptans* (ajuga):** Striking variegations and colors

***Alchemilla mollis* (lady's mantle):** Chartreuse color; shaped like a maple leaf

***Artemisia* spp. (artemisias):** Silver or green; aromatic; fernlike

***Bergenia* spp. (bergenias):** Evergreen but burgundy in fall; glossy

***Hosta* spp. (hostas):** Many solid and variegated colors; smooth or puckered surface

***Polygonatum odoratum* (Solomon's seal):** Long, graceful shoots; green or variegated

***Pulmonaria* spp. (lungworts):** Dark green or variegated

***Sedum* spp. (sedums):** Many colors; variegated forms; fleshy

***Stachys byzantina* (lamb's-ears):** Silver-gray; velvety

***Yucca* spp. (yuccas):** Evergreen; large and sharply pointed

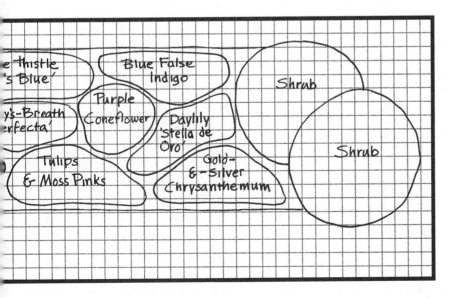

e thistle 's Blue'

Blue False Indigo

Shrub

Purple Coneflower

y's-Breath erfecta'

Daylily 'Stella de Oro'

Shrub

Tulips & Moss Pinks

Gold- & Silver Chrysanthemum

Shrub

Garden Plotter

GARDENER TO GARDENER

Instant Garden Beds

We had to move frequently over the last few years, and the ground around our rented homes often was nothing more than thick sod over hard clay. Yet we were able to create a fertile growing site quickly and easily. Here's how:

Buy several bags of commercial topsoil and cut 8 to 10 drainage holes on one side of each bag. Lay the bags (holes on the bottom) over the area where you want to garden. Cut a large X from corner to corner on each bag.

Water the exposed topsoil to flush away any salts, let it drain for a couple of days, then plant and mulch right in the soil on top of the bag.

The following spring, remove the plastic from underneath the soil—the sod will be decayed. Use a spade to turn the soil and form a regular raised bed.

Kit Kellison,
Chesapeake, Virginia

Christmas Tree Trellises

I recycle my neighbors' discarded Christmas trees into trellises for my pole beans. I gather the trees in a flat spot near my garden, then use a chainsaw or lopping shears to remove all of the branches except a few near the top. (I run the branches through a chipper/shredder and use the resulting mulch around my acid-loving rhododendrons and azaleas.)

In spring, when it's time to plant my pole beans, I erect my trellises. For each trellis, I gather three trimmed trees, teepee style, interlocking the small branches at the top. For extra support, I wrap string all around the sides of each trellis.

Fred Robitzer
Reading, Pennsylvania

Space-Saving Combo

Interplanting two crops in one space is a great way to boost your vegetable yields. My favorite crop combo is early peas and sweet potatoes. First I plant my early peas at the recommended spacing. Then, after all danger of frost has passed, I plant the sweet potatoes between the rows of growing peas. When the pea plants die back, they become mulch for the sweet potatoes, and I get two crops from the same amount of space.

Carol Hershberger
Hutchinson, Kansas

Easy Garden Start-Up

We reclaimed a garden site that had long ago been taken over by weeds, and we did it with hardly any work. After an early spring rain, we rolled out a 20-foot-wide length of black plastic and covered the bed. Then we waited. Within 6 weeks, the weeds were smothered and worms were working in the warmed soil. We cut crosses in the plastic at regular intervals, folded back the corners, worked in some compost, and then planted seedlings. The plants grew quickly and produced bountifully with little watering and no weeding.

Nancy Prentiss
Farmington, Maine

Help for Transplanted Gardeners

When you move to a new region, seek out a local nursery to help you adjust to the unfamiliar gardening conditions. When I moved from Connecticut to Arizona 3 years ago, a tiny local nursery helped me learn how to plant the right kinds of plants at the right times for our area.

Jack McGarvey
Rio Rico, Arizona

Chapter 2

February

Warming Up: Starting Seeds and Caring for Seedlings

Now draw the plan of our garden beds,
And outline the borders and the paths correctly.
We will scatter little words upon the paper,
Like seeds about to be planted.

—Amy Lowell

Ready, set, grow! For many gardeners, February marks the beginning of planting season—time to plant seeds of many vegetables and flowers indoors for transplanting to the garden later this spring. (Gardeners in the most southern climates, already in high gear, began planting outdoors last month.)

Starting plants from seed is one of the most satisfying garden activities. Besides saving you money, starting from seed gives you access to hundreds of cultivars not otherwise available. But there's another, very special reason to start your own plants from seed: the experience! Even the most expert seed starters continue to feel awed by the miracle of germination and growth—how one seed (sometimes not much bigger than a speck of dust) can become a healthy, productive plant that, in some cases, is large enough to "take over" the entire garden! By starting your own seeds, you can help make it happen and ensure that your garden is 100 percent organic, right from the start.

If you haven't started plants from seed before, make this the year you begin. We'll tell you how to provide the conditions that seeds need to germinate, as well as how to get your seedlings off to a strong, organic start.

Gardener's To-Do List—February

**If you don't know what USDA hardiness zone you live in,
check the map on page 230 to find out.**

Zone 3

- [ ] Garden indoors! Grow some leaf lettuce, chives, and cress beneath lights.
- [ ] Move geraniums that have been stored indoors for winter into brighter light. Slowly coax them back to life with dribbles of water and weak fertilizer.
- [ ] Late in the month, sow onion, celery, pansy, snapdragon, and viola seeds indoors.

Zone 4

- [ ] Sow seeds of onions and chives indoors under lights.
- [ ] Late in the month, start some seeds indoors of early cabbage, broccoli, cauliflower, and celery, as well as of pansies, daisies, and other hardy flowers.
- [ ] Prepare your coldframe for use next month.
- [ ] Check peonies and other perennials for signs of heaving. Press them back into place with your foot if their roots show aboveground.
- [ ] Lightly trim damaged wood from mature apple trees, but postpone heavy pruning for a few more weeks.

Zone 5

- [ ] Kick off the growing season: Start celery, onions, leeks, and hardy herbs indoors beneath lights. Toward month's end, start early cabbage.
- [ ] Outdoors, cover compost with black plastic to help it thaw.

- [ ] Start cold-hardy flowers, such as snapdragons, stocks, pansies, and lobelias, indoors beneath grow lights.
- [ ] Between snows, direct-seed poppies and larkspur outdoors.
- [ ] Order new disease-resistant apple trees such as 'Enterprise', 'Freedom', and 'Goldrush'.

Zone 6

- [ ] Indoors beneath lights, start cabbage, cauliflower, celery, and broccoli.
- [ ] Outdoors, work the soil as soon as it has dried, preparing beds for peas, potatoes, and other early crops.
- [ ] Check stored dahlias and pot up those that have begun to sprout.
- [ ] Prune back clematis vines, and sow Shirley poppies outdoors wherever you want them to bloom.
- [ ] Start seeds of slow-growing petunias indoors beneath lights.
- [ ] Root geranium cuttings.
- [ ] Late this month, cut back ornamental grasses.
- [ ] Prune apple trees and rake the orchard floor to interrupt the life cycles of insect pests.

Zone 7

- [ ] Start slow-growing lettuce and early tomatoes indoors under lights.
- [ ] Set seed potatoes in a warm place to encourage sprouting.
- [ ] Harden-off cabbage seedlings outdoors in a coldframe; toward month's end, plant them in the garden beneath cloches.
- [ ] Outdoors, sow peas and parsley near the end of the month.
- [ ] Mow winter cover crops and turn them under if the soil is dry enough to cultivate.
- [ ] Spread compost over beds that you will plant next month.
- [ ] Late in the month, prune hybrid tea roses hard, but give shrub roses only a light trim.
- [ ] Weed strawberries and blanket them with a row cover to encourage early bloom.

Zone 8

- [ ] Harvest asparagus spears all month!
- [ ] Plant potatoes and mulch them well.
- [ ] Pop onion sets or plants into fertile, well-drained beds.
- [ ] Late this month, direct-seed lettuce, endive, and other leafy greens in the garden, and prepare a bed for carrots.
- [ ] Start tomatoes and peppers indoors; as soon as the seeds sprout, move the flats or pots beneath bright lights.
- [ ] Weed and clean the garden beds where your pansies and other hardy spring flowers will soon bloom.

- [ ] Set out new roses, and heavily prune established hybrid tea roses.
- [ ] Expand your selection of small fruits! Plant figs, blackberries, blueberries, and strawberries. Also fertilize established fruits.

Zone 9

- [ ] Finish planting cabbage, celery, peas, and broccoli early this month.
- [ ] By midmonth, direct-seed beets, carrots, and leaf lettuce.
- [ ] Mulch potato plants.
- [ ] Thin leafy greens and water them regularly.
- [ ] Late this month, begin planting beans, cucumbers, and early sweet corn. Also sow annual herbs, such as dill, fennel, and basil.
- [ ] Indoors, be sure to keep tomato, pepper, and eggplant seedlings beneath bright lights.

Zone 10

- [ ] Sow bachelor's buttons, nasturtiums, and California poppies outdoors.
- [ ] Cut back ornamental grasses to the point where you see new growth.
- [ ] Plant muscadine grapes and figs. Mulch them well to keep the roots evenly moist.
- [ ] Control pest caterpillars on vegetables with *Bacillus thuringiensis* (BT).

Seed Starting Simplified

If you're like many gardeners, you have never tried growing your own plants from seed. Or, if you have tried, maybe your seedlings didn't resemble those you see at the garden center each spring, and you're wondering how you can do better.

Rest assured, starting your own seedlings is fun, easy, and well worthwhile. By growing your own transplants, you can choose from hundreds of unusual varieties—including those with tolerance to heat or cold, disease resistance, and unmatched flavor—that simply aren't available at garden centers. Plus, you'll have the satisfaction of knowing that you've grown your entire garden organically right from the very start.

For seed-starting success, follow this simple plan.

1. Choose a fine medium. For healthy seedlings, you've got to give them a loose, well-drained *medium* (seed-starting mix) composed of very fine particles. You can buy a seed-starting mix at your local garden center. Don't use *potting soil*—often, it's too rich and doesn't drain well enough for seedlings.

2. Assemble your containers. Many gardeners start their seeds in leftover plastic "six packs" from the garden center, empty milk cartons, or Styrofoam cups. If you don't have containers on hand, you can buy plastic "cell packs," individual plastic pots, or sphagnum peat pots. Or make your own pots from newspaper (see page 48). Whatever you use, be sure your containers drain well (usually through holes in the bottoms of the containers).

Set the pots inside a tray so that you can water your seedlings from the bottom (by adding water to the tray) rather than disturbing them by watering from the top. You can buy seed-starting trays at garden centers and many hardware stores.

3. Start your seeds! Moisten your seed-starting mix *before* you plant your seeds. If you water after you plant the seeds, they can easily float to the edges of the container—not where you want them to be. To moisten the mix, simply pour some into a bucket, add warm water, and stir. After about 8 hours (or when the mix has absorbed the water), fill your containers with the moistened mix.

Plant at least two, but no more than three, seeds per container. The seed packet usually tells you how deep to plant, but a good rule of thumb is three times as deep as the seeds' smallest diameter. (Some flower seeds require light to sprout—if that's the case, simply lay the seeds on the surface of the mix, then tamp them in gently with your finger.)

(see page 48)

Why Start Your Own Seeds?

Sure, it's easy to pick up a few "six packs" of tomatoes and marigolds at the garden center. But why do that, when you can have . . .

◆ **Selection, selection, selection.** Nurseries tend to stock only tried-and-true varieties. If you start your own seeds, the selection is virtually limitless.

◆ **Earlier harvests and blooms.** You can jump-start the season and enjoy fresh food and flowers sooner.

◆ **More for less.** Even after you figure the cost of supplies, growing your own transplants is still cheap. At the garden center, a flat of 36 petunias, for instance, will cost you about $12, but a packet of 100 seeds costs only $2 or so.

◆ **Strong, healthy transplants.** Who knows how long those store-bought seedlings languished in the greenhouse? By growing your own plants, you control quality.

◆ **Organic guaranteed.** No need to worry about chemicals. You can make sure your garden is organic from seed to fruit and flower.

After you've planted your seeds, cover the tray loosely with plastic to create a humid environment. At 65° to 70°F, your seeds should sprout just fine without any supplementary heat. If the room temperature is cooler than that, you can keep the seeds warm by setting the tray on top of a heating mat made specifically for starting seeds.

Tomato, zucchini, and pumpkin seeds should push their sprouts through the surface of the mix in a few days. Peppers sprout in about a week. And some seeds, such as parsley, can take as long as 3 weeks to sprout, so be patient.

4. Keep the lights bright. Check your trays daily. As soon as you see sprouts, remove the plastic covers and immediately pop the trays beneath lights. You can invest in grow lights (which provide both "warm" and "cool" light), but many gardeners have good results with standard 4-foot-long fluorescent shop lights. Set your seedlings as close to the light as possible—2 or 3 inches away is about right. When seedlings don't get enough light, they grow long, weak stems. As the seedlings grow, raise the lights to maintain the proper distance.

And don't worry about turning off the lights at night. Contrary to popular belief, seedlings don't require a period of darkness. Fluorescent lights are only one-tenth as bright as sunlight, so your seedlings will actually grow better if you leave them on continuously.

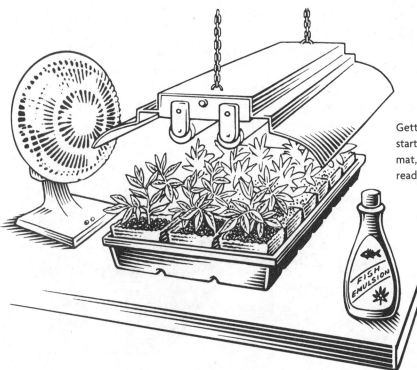

Getting started! The seeds, seed-starting mix, containers, tray, heating mat, and fan are assembled and ready to go.

When to Start Your Seeds

To calculate when to sow your seeds, you first need to know your last expected frost date. (If you don't know, call the Cooperative Extension Service.) When you have your last frost date, get a calendar and count backward from it, using the seed-to-transplant time given on the seed packet or catalog.

If you're starting crops for fall harvest, start with the date at the *other* end of the growing season: the average first frost date. To squeeze in a crop before the first fall frost, count backward the number of days listed on the seed packet or in the catalog. Then, start your seeds a week or two *before* that date to compensate for plants' slower growth late in the season.

5. *Feed and water.* Your seedlings will need a steady supply of water, but the soil shouldn't be constantly wet. The best method is to keep the containers inside a tray, water from the bottom, and allow the soil inside the containers to "wick up" the water.

If your growing medium contains only vermiculite and peat (as many seed-starting mixes do), you'll also need to feed your seedlings. When the seedlings get their first "true" leaves (not the tiny ones that first appear, but the two that follow), mix up a fish emulsion solution one-quarter to one-half the recommended strength and add it to the seedlings' water every other week. As the plants grow bigger, gradually increase the strength of the mixture.

6. *Keep the air moving.* Your seedlings need to be big and strong by the time you move them from their cushy indoor surroundings to the harsh realities of the outside world. You can help them grow sturdy, stocky stems with a small fan. As soon as you see those first true leaves, set the fan to blow lightly but steadily on the seedlings, all day long. The air circulation also will minimize their chance of fungal disease while they're crowded together indoors.

7. *Give them space.* Those well-watered, well-fed, and well-fanned seedlings will soon need more root space. Shortly after the second set of true leaves appears, take a deep breath and thin your seedlings to one per pot. Use small scissors to clip off the weaker plants at the soil line, leaving only the stockiest plant.

Next, carefully "pot up" the survivors into larger, 3- or 4-inch pots. Squeeze the sides of the smaller containers all around, turn them upside down, and the plants should come out easily—soil and all. Immediately set them into the larger containers and fill with a mixture of 3 parts potting soil and 1 part your own screened compost.

Crowded seedlings will need to be thinned, or they'll produce fewer flowers and less of a crop. Snip off stems at soil level, being careful to avoid disturbing the roots.

(If you started your seeds in peat pots or homemade newspaper pots, you can plant both the seedling and its pot in the larger container; the pot eventually will decompose.)

Plant tomatoes deep in the new container to encourage them to develop a larger root system to support these often top-heavy plants. With most other plants, the soil level in the new pot should be about the same as in the smaller container. After you've finished repotting, water the plants well and set them back under the lights.

8. *Harden-off.*

About a week or two before you plan to transplant your seedlings to the garden, begin taking them outdoors to a protected place, such as inside a coldframe or near a wall, for increasing lengths of time on mild days. This will help them adjust to the conditions outside—a process known as *hardening off*. Start with just a couple of hours each day, work up to a full day, and then leave them out overnight.

When you finally transplant the seedlings to the garden, be careful not to disturb their roots. Gently pop them out of their containers, keeping as much soil attached to their roots as possible. Again, plant tomatoes deeply, but set other plants at about the same depth as they were in their pots (or just slightly deeper).

9. *Seal it with a K.I.S.S.*

Most important, relax! Don't worry if you forget to do something or don't follow all the "rules." Except for hardening off, all of these rules are flexible. Before long, you will learn what works best for you—and will have a few secrets of your own to share with fellow seed starters.

Preparing for moving day: Plants need a week or two of "hardening off" to help them adjust to outdoor conditions.

How to Read a Seed Packet

Most seed packets contain useful information that can help you select the best seeds for your needs and get them off to a strong start. But to put that information to use, you need to know how to read it! Here's a guide to seed-packet shorthand.

Certified organically grown
When you see this label on a packet, it means the seeds were grown in accordance with the standards of an organic certification organization.

Planting information
Here, you can find advice on how deep to plant the seeds, how long it will take them to sprout, and what the seedlings will look like. Plus, you can figure out how many seeds you'll need in order to plant your garden.

Seed count
Some packets have seed counts (as well as weight) to help you figure out how many packets you need to buy.

These seeds are certified organically grown in accordance with Oregon Tilth standards and meet or exceed Federal germination requirements. They are unconditionally guaranteed or your money back.

Hollyhock
Summer Fantasy Mixed Colors

Contains about 50 seeds.

Tall, stately stalks topped by fully double flowers in various shades of peach, maroon, mauve, fuchsia, pink, and red give dramatic impact to your garden and foundation plantings.

Packet Plants	Plant Spacing	Planting Depth	Days to Germination	Seedling Identification
60 ft.	24 in.	¼ in.	10-14	

Growing Tips:
After soil warms in spring, seed directly into the garden in sunny location. For transplanting, start indoors in sunny window 6-8 weeks before transplant date. Handle transplants with care. In unprotected areas, stake tall plants for support.

Uses:
Add an old-fashioned look to fences and walls.

When to Plant Outdoors

JUNE
MAY-JUNE
MAR-MAY
FEB-APR
FEB-MAR

Garden Plan

Hollyhock
Gypsophila   Tall Poppy
Melampodium

Packed for 2001

Uses
Flower packets sometimes tell you how to use particular cultivars in the landscape. Vegetable packets may include cooking tips or recipes.

Growing tips
Here you'll find details on where to plant the seeds, how to grow seedlings indoors, and how to care for the plants once they're planted in the garden.

Planting map
Figure out where you are on the map, then refer to the color guide to determine the prime time for planting outdoors in your area.

Garden plans
Packets may have plans as a visual aid for designing flowerbeds and gardens.

Packet date
The packet is marked with the year for which the seeds were tested for minimum germination. But don't be afraid to take a chance on slightly older seeds that often go on sale at the end of the season. Most seeds (with the exception of a very few, such as onions, parsley, and parsnips) will sprout just fine if they're a little old. Avoid packets with water stains, though; seeds that have been exposed to wet conditions probably won't grow well. Store your bargain seeds at home in a cool, dry place. Plant them a little more thickly as extra insurance.

Seed-Starting Equipment and Supplies

Gardeners seem to fall into two different camps when it comes to seed-starting equipment. Some take a freewheeling approach, swearing by homemade setups that involve yogurt cups, shop lights, and old ironing boards. Others are more methodical and will use nothing less than highly sophisticated grow lights, heating cables, and special propagation units purchased from catalogs or garden centers.

You may find that the best approach is a combination of the two. While it's true you don't *need* a lot of fancy equipment to start your own seeds indoors, having at least a few gadgets designed specifically for starting seeds can make the process easier and more fun.

Containers

You can start seeds in almost any kind of container that will hold 1 to 2 inches of starting medium without becoming waterlogged. After seedlings form more roots and develop leaves, though, they grow best in larger individual containers that provide more space for root growth and have holes for drainage.

Flats are large, rectangular containers that hold many seedlings. Many gardeners start their seeds in them, then transplant the seedlings to individual containers after the first true leaves unfold. If you raise lots of seedlings, it's useful to have interchangeable standard-size flats and inserts. You can buy flats at a garden center, or make your own by constructing a rectangular wooden frame, 3 to 4 inches deep. Nail slats across the bottom, leaving ⅛ to ¼ inch between them for drainage.

Although **individual containers** dry out faster than flats, they are better for starting seeds because you don't have to repot as often, so the seedlings' tender roots are less likely to be damaged by constant handling. Some containers, such as peat pots, paper pots, and soil blocks, go right into the garden with the plant during transplanting so the plants' roots are never disturbed.

If you choose to use homemade containers, such as old **milk cartons, yogurt cups,** or **egg cartons**, keep in mind that square or rectangular containers make better use of space and provide more area for roots than round ones do. Also, be sure to poke a drainage hole in the bottom of each container.

Or you could splurge: Spend a couple of bucks on containers that are *designed* for starting seeds. Most garden centers and many home and hardware stores carry **cell packs,** plastic trays that have individual 2- or

3-inch-deep pockets with drainage holes. (Leftover "six-pack" containers from the garden center will work fine, too.) These special-purpose packs range from 6 cells to 40 or more, and many include a clear plastic dome that helps maintain humidity until the seeds have sprouted.

Peat pots, made entirely of peat moss, are popular because you can plant them "pot" and all—you don't have to worry about extracting the seedlings from the containers before you set them in the garden. Also, the peat absorbs excess moisture naturally, so seedlings are less susceptible to damping-off, a fungal disease that often occurs when soil is too soggy. But because peat pots do dry out faster than plastic containers, you'll need to check their moisture level daily.

Like peat, **paper pots** also break down in the soil, allowing you to place them right in the garden, pot and all. They also draw excess water away from the seed-starting medium, although not to the degree that peat does. You can buy pots made from recycled paper or make your own pots from newspaper strips. (See page 48.)

Another option is to skip the pots completely and start your seeds in **soil blocks.** If you go this route, you can do without containers, but you will need a **soil-block maker**—a device that compresses the seed-starting medium into cubes in which you plant your seeds. You can choose between block makers that produce four 1¾-inch cubes (just right for big seeds, like those of tomatoes, peppers, and cucumbers) or twenty ¾-inch cubes (for lettuce, alyssum, and other tiny seeds). Like peat and

The tools for starting seeds are easy to assemble. A bag of growing medium, water, and simple or fancy containers will get you off on the right foot.

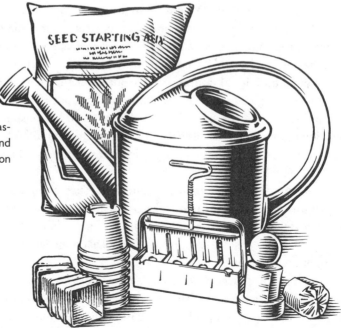

paper pots, soil blocks go right into the garden without disturbing delicate seedling roots. But, also like peat, the cubes dry out quickly, so you'll need to water them often to keep the seedlings from drying out. See page 29 for details on making soil blocks.

No matter what kind of container you start your seeds in, you'll also need a **tray** to put beneath the containers. This allows you to water your seedlings by filling the tray rather than dumping water on them from the top. A **plastic lid** or **wrap** for covering the containers after you seed them will help keep the seed-starting mix moist and encourage germination. Also be sure to have some simple **markers** or labels on hand so that you can note the variety and sowing date.

Seed-Starting and Potting Mixes

Seeds contain enough nutrients to nourish themselves, so a seed-starting mix doesn't have to contain nutrients. But it should provide plenty of air spaces, hold moisture well, and be free of weed seeds and toxic substances. **Peat moss, compost, perlite,** and **milled sphagnum moss**—either alone or in combination—are all good materials for starting seeds. Don't use plain garden soil, though; it hardens into a dense mass that seedling roots can't penetrate.

When your seedlings have their first set of true leaves, you'll need to transplant them to a **nutrient-rich potting mix**. You can either use a commercial mix (check the label to make sure it doesn't contain a synthetic chemical fertilizer) or make your own. To make a basic mix, try this popular organic potting soil recipe:

- 1 to 2 parts good-quality garden soil

- 1 part builder's sand or perlite

- 1 part compost

Each component provides specific benefits to plants. Soil contains essential minerals. Sand and perlite assure good drainage. (Perlite, an expanded volcanic rock with many air spaces, will make the mix lighter than if you use sand.) And compost releases nutrients slowly, helps maintain proper soil pH, improves drainage, and holds moisture.

Lights

Seedlings need more intense light than full-grown plants. If they don't get 14 to 16 hours of strong light a day, most become spindly and weak. Although many gardeners start their seeds on windowsills, the light from a window during the short days of winter often isn't enough to grow strong, sturdy seedlings.

Light Advice

A grow-light system will provide seedlings started indoors with enough light to produce healthy, compact transplants for the garden. Here's how to get the most from your lights:

- Growing seedlings need lots of bright light. Keep the lights on for at least 14 hours a day, and suspend the lights close to seedling leaves.

- Because tubes produce less light at the ends, choose the longest tubes you have room for and rotate seedlings at the ends into the middle every few days.

- Keep the tubes clean. Dust can decrease the amount of light available.

- To increase light from a fluorescent fixture, position a mirror or aluminum foil alongside it to reflect light back onto seedlings.

Most seedlings will do better if you grow them beneath fluorescent lights. You can buy expensive grow lights, but the 4-foot-long shop lights sold at hardware stores work just as well and cost much less. Start with new tubes—fluorescent tubes become dimmer over time. Don't bother with incandescent bulbs. Their light does not stimulate growth well.

Suspend your light fixture from the ceiling over a table or bench. To protect the table or bench from water, cover it with a plastic sheet.

Heating Mats and Cables

Most seeds—including tomatoes and peppers—germinate much faster in warm soil (about 70° to 75°F). To provide those toasty temps, heating mats and cables come in handy. You just plug them in and set your containers on top of them.

Some of these devices connect to a control unit that allows you to set the temperature at the exact level needed for germination. Others have thermostats, switching on and off automatically to keep the soil temperature in the mid 70s or so.

Keep in mind that most sprouted seedlings grow better in slightly *cooler* temps (upper 50s to lower 60s), so remove the heating mat or cables after the seeds have sprouted.

Capillary Mats

For peat pots, paper pots, and soil blocks, a capillary mat can be a big help. Capillary mats allow the seed-starting medium to draw water from a reservoir as needed, so you don't have to monitor moisture levels as often. Simply set your containers on top of the mat and keep the reservoir filled (usually one filling every 4 or 5 days is all that's needed). The mats are made of a fiber that "wicks" water, spreading the moisture evenly from the reservoir to all corners of the mat. You can even put liquid fertilizer in the reservoir to feed larger seedlings. One drawback is that, as the seedlings grow, their roots may attach to the mat; if that happens, just peel off the mat at transplant time.

Fertilizer

Seedlings growing in a soil-free or lean potting mix will need small doses of plant food, starting at the time the first true leaves develop. For the first 3 weeks, water them once a week with a half-strength solution of fish or seaweed fertilizer, compost tea, or one of the liquid organic fertilizers specially formulated for seedlings. After that, feed the seedlings with a normal-strength solution every 10 to 14 days. If you're growing your seedlings in a potting mix that contains compost or other nutrients, you may not need to feed them as often.

Technique

How to Make Soil Blocks

Compared to yogurt containers, foam coffee cups, and milk cartons, using soil blocks for seed-starting offers some real advantages that easily justify the cost of a soil-block maker (see page 26). With these compressed freestanding cubes, you need no other containers. And you plant the blocks directly into the garden, with little or no setback to the plant. What's more, making soil blocks is fun. Here's how:

1. To make the blocks, you'll first need to mix up a moisture-retaining "soil." In a wheelbarrow or large bucket, combine 3 parts peat, 2 parts coarse sand or perlite, 1 part good-quality garden soil, and 2 parts screened, well-aged compost. Gradually add water to the mix until it reaches the consistency of cake batter.

2. Fill the compartments of the soil-block maker with wet soil mix. Press the bottom of the block maker onto a hard surface to fill it completely and force away any excess mix.

3. Position the block maker in the flat where the seedlings will grow. Squeeze the handle as you lift the block maker up and away to press out the blocks. Continue pressing out blocks until the flat is filled. If the blocks stick, dip the block maker in water and try again.

4. Sow one or two seeds in each of the small indentations on top of the cubes. Cover the entire flat with plastic wrap, a polyester row cover, or a similar material to maintain humidity. Remove the cover as soon as the seeds germinate.

Watering Soil Blocks

Soil blocks dry out more quickly than plastic pots or wooden flats, so you'll need to check them daily and be prepared to water as soon as the blocks show any sign of dryness.

Water the blocks gently. Mist them with a sprinkler or use a spray wand that delivers very fine droplets of water—never spray them directly with a hose. Capillary mats, which deliver water as needed, are also excellent for soil blocks. Instead of watering the blocks every day or two, just fill the reservoir below the capillary mat—usually only once every 4 or 5 days. Within a few weeks, fast-growing seedlings, such as lettuce, will be ready to transplant to the garden.

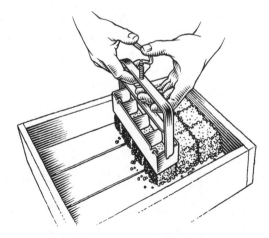

Press out soil blocks until the flat is filled.

Sprinkle a seed or two in each soil block.

Seed Inventory

From season to season it's all too easy to lose track of the seed packets you've ordered and haven't entirely used up. If you store the leftovers from one season's planting in a cool, dry place, most seeds will remain usable for up to 3 years. This worksheet should help you keep track of the odds and ends of various vegetable, herb, and flower seeds you have on hand. Remember to test stored seeds before planting to make sure they'll still germinate.

Cultivar	Date Purchased	Amount Purchased	Seed Source	Amount Left After Sowing	To Be Used By	Amount to Reorder

Cultivar	Date Purchased	Amount Purchased	Seed Source	Amount Left after Sowing	To Be Used By	Amount to Reorder

Place Sure Bets

Some plants are easier to start from seed than others. If you've never tried starting your own plants from seed, you might want to stick with these surefire vegetables and flowers.

VEGETABLES
Broccoli
Brussels sprouts
Cabbage
Cauliflower
Leeks
Lettuce
Onions
Peppers
Tomatoes

FLOWERS
Alyssum
Cosmos
Marigolds
Shasta daisies
Zinnias

Guide to Starting Vegetable Seeds

By starting seeds of the following crops indoors, you can get a head start on the growing season, escape poor outdoor germination conditions, and try rare and unusual varieties. (Note that some plants—such as root crops, beans, and corn—grow best when seeds are planted directly in the garden. For more about direct seeding, see Chapter 4.)

Broccoli *(*Brassica oleracea, *Botrytis group)*

Delicious either cooked or eaten raw, broccoli freezes well and contains large amounts of nutrients, such as vitamins A and C, as well as B vitamins, calcium, and iron.

To start seeds indoors: Start spring crops in individual pots 7 to 9 weeks before the last expected frost date. Plant seeds ¼ inch deep in seed-starting mix. Keep the soil moist, but not wet. The seeds should germinate in 5 to 10 days. Transplant the seedlings into the garden when they are about 6 inches tall, with two to four leaves. (*Note:* Direct-seed fall crops in midsummer.)

Brussels Sprouts *(*B. oleracea, *Gemmifera group)*

Brussels sprouts yield a large harvest from a small space. The mini-cabbages form along the stems beneath umbrella-like foliage and need up to 100 days to mature.

To start seeds indoors: Brussels sprouts produce best in cool weather, so check your seed packet for the days to maturity, then count back that number of days from your first expected fall frost. Start your seeds 6 to 8 weeks before that date. Sow seeds ½ inch deep in individual pots; germination should occur in 5 to 10 days. When the seedlings are hardened off and ready to go out into the garden, set them deeper than they grew originally, with the lower leaves just above the soil. (*Note:* Direct-seed fall crops in midsummer.)

Cabbage *(*B. oleracea, *Capitata group)*

In most areas, you can plant an early crop for fresh eating and a late crop for storage. Choose from early-, mid-, and late-season cultivars with green, red, or purple heads.

To start seeds indoors: Plan to set out early types 4 to 5 weeks before the last expected frost. Sow seeds ¼ inch deep and 2 inches apart in flats, about 10 to 12 weeks before the last expected frost date. When seedlings have three leaves and daytime temperatures reach 50°F, plant them outdoors slightly deeper than they grew in their flats. (*Note:* Mid- and late-season types can be either direct-seeded in midsummer or started indoors.)

Cauliflower (B. oleracea, *Botrytis group*)

Primarily a cool-weather crop, cauliflower won't produce heads if weather turns hot. Most cultivars need about 2 months of cool weather to reach maturity.

To start seeds indoors: Plant seeds ¼ to ½ inch deep, 8 to 10 weeks before the last frost date. Set the pots in bright light and provide even moisture. Use bottom heat, if necessary, to keep the soil temperature around 70°F until the seeds sprout; remove the heat source once seeds have germinated. About 2 to 4 weeks before the last expected frost, transplant seedlings to the garden. (*Note:* Direct-seed fall crops about 2 to 3 months before the first expected frost.)

Celery (Apium graveolens *var.* dulce)

Contrary to popular belief, celery isn't that difficult to grow if you give it rich soil, plenty of water, and protection from high temperatures.

To start seeds indoors: For a late-summer crop, sow seeds 8 to 10 weeks before the last average frost. Soak the tiny seeds overnight to encourage germination. Fill flats with a mixture of moist seed-starting mix and 20 to 30 percent well-aged compost, and plant the seeds in rows 1 inch apart. Cover them lightly with more seed-starting mix, then cover the flats with damp sphagnum moss or burlap until the seeds sprout. Keep flats at a temperature of 70° to 75°F during the day and about 60°F at night. When seedlings are about 4 inches tall, transplant them into individual peat or newspaper pots. When they are 6 inches tall, harden them off, then transplant them to the garden. Keep plants well fed and watered until harvest time.

Eggplant (Solanum melongena *var.* esculentum)

Eggplants produce glossy fruits in a variety of colors, shapes, tastes, and sizes. Most cultivars need 100 to 150 frost-free days to produce a bountiful crop.

To start seeds indoors: Start seeds 6 to 8 weeks before the average last frost. Sow seeds ¼ inch deep and use bottom heat to maintain a soil temperature of about 80°F for 10 days or so, until they germinate. Once germination occurs, remove the heat source. When seedlings are 3 inches tall, transplant them to individual peat or newspaper pots. Seedlings can be transplanted outdoors when nighttime temperatures reach at least 50°F and the soil warms to at least 70°F.

The AAS Seal of Approval

Ever wonder about the All-America Selections winner symbol that appears on seed packets and labels and in garden catalogs? It indicates that a nonprofit organization, All-America Selections (AAS), has tested the plant and selected it as a superior cultivar for home gardens.

AAS tests new cultivars each year at dozens of sites across North America, including universities, botanical gardens, and other horticultural facilities. New entries are compared to past winners and standard cultivars. Vegetables are evaluated for traits such as yield, flavor, pest resistance, and space efficiency; flowers are judged for long-lasting blooms, uniformity, and disease resistance. Only about 5 percent of the cultivars entered each year will win awards.

For more information, write to All-America Selections, 1311 Butterfield Road, Suite 310, Downers Grove, IL 60515; or visit their Web site at www.all americaselections.org.

Leek (Allium ampeloprasum)

The sweetest and most delicately flavored of all onions, leeks are easy to grow but require 70 to 110 days to mature.

To start seeds indoors: Sow seeds ¼ inch deep and 1 inch apart in flats, 2 or 3 months before the last average frost date. Keep the flats at a temperature of 65° to 70°F during the day and 55° to 60°F at night. When seedlings are about 8 inches tall, transplant them to the garden, planting them in 6-inch-deep drills so that the tips are just a few inches above the surface. Don't fill in the drills—allow rain or watering to wash the soil back in for you.

Lettuce (Lactuca sativa)

Lettuce greens are easy to grow and demand little space. You can sow lettuce seed directly in the garden, but by starting successive crops indoors, you can harvest tasty homegrown lettuce almost year-round.

To start seeds indoors: Sow seeds 4 to 6 weeks before the last frost date, ⅛ to ¼ inch deep and 1 inch apart, making three small sowings at weekly intervals. Seeds will sprout in 7 to 10 days. Set out the seedlings successively as soon as the ground is workable. Repeat the procedure in midsummer for a fall harvest that can last until frost.

Melon (Cucumis melo *and other genera*)

Most melons need nutrient-rich soil, plenty of sunshine, and at least 3 or 4 months of warm weather. But the superior taste of vine-ripened melons makes them well worth growing.

To start seeds indoors: Plant several seeds ½ inch deep in each 4-inch peat pot or soil block, just 2 to 4 weeks before transplanting—seedlings that develop tendrils or more than four leaves before they are transplanted may have trouble establishing roots. If necessary, provide bottom heat to bring the soil temperature to about 80°F. Thin 2-inch seedlings, leaving only the strongest plant in each pot. Transplant them, pot and all, when all danger of frost has passed and garden soil has warmed to at least 70°F. Make certain that the top of the peat pot is well below the soil line; otherwise, it will act as a wick and dry out the pot, making it difficult for tender young roots to break through. Some gardeners peel off an inch or so of the rim as added insurance.

Onion (Allium cepa *and other species*)

Onions are easy to grow, healthful, and indispensable for cooking. Although you can grow them from sets or plants, growing onions from seed gives you a much greater choice of cultivars.

To start seeds indoors: Sow the seeds in flats about 10 to 12 weeks before the last expected frost, ¼ inch deep and four seeds per inch. Seeds will sprout in about 10 to 14 days. If the seedlings start getting tall and gangly before you're ready to plant them, cut them back a couple of inches to encourage them to become more stocky. Transplant seedlings to the garden, starting 6 weeks before the last frost date.

Pepper (Capsicum annuum *var.* annuum)

Pepper choices cover an amazing range of flavors, shapes, textures, colors, and sizes. But to grow the most intriguing cultivars, you must start from seed.

To start seeds indoors: Plant seeds 2 months before the last frost date. Sow two or three seeds per pot, ¼ inch deep. Maintain a soil temperature of about 75°F, and keep the soil moist but not wet. When seedlings are 2 or 3 inches tall, thin to the strongest plant in each pot. Transplant 4- to 6-inch seedlings to the garden when the soil temperature is at least 60°F, usually 2 or 3 weeks after the last frost.

Squash (Cucurbita *spp.*)

Summer squash, such as zucchini, produces prolifically throughout the summer. Winter squash, which includes acorn, butternut, and buttercup, typically takes longer to mature. Both types need warm weather, lots of sun, and plenty of room.

To start seeds indoors: In areas with short growing seasons, sow seeds about a month before the last expected frost; place two seeds ½ to 1 inch deep in each peat pot or soil block. After the seedlings emerge, thin to just the strongest seedling. Water well just before transplanting outdoors—about a week after the last frost date, when the soil temperature has warmed to at least 60°F.

Tomato (Lycopersicon esculentum)

America's favorite vegetable, the tomato, comes in hundreds of different cultivars, including beautiful heirloom types, early-maturing types, and those bred for special climates and for disease resistance. For the largest selection, start your tomatoes from seed.

To start seeds indoors: Sow seeds ¼ inch deep either in cell packs or about 1 inch apart in flats, 6 to 8 weeks before the average last frost. Seeds will germinate in about a week if the soil temperature is 75° to 80°F. After the seedlings emerge, keep the temperature no higher than 70°F. Feed once a week with diluted fish emulsion. When the first true leaves appear, transplant seedlings to individual 4-inch peat pots, burying the stems slightly deeper than they were before. After all danger of frost has passed, transplant the seedlings to the garden, again burying the stems deeper than they were in the pots.

Presprouting Saves Time

You can get peas, corn, melons, and squash off to a fast start (and earlier harvest) by presprouting the seeds indoors. Here's what to do:

1. Spread a double layer of damp paper towels on a flat surface. Evenly space the seeds about 1 inch apart on the towels.

2. Carefully roll up the towels without disturbing the seeds; put a rubber band around each roll.

3. Enclose one or more rolls in a plastic bag. Close the bag loosely—germinating seeds need some air.

4. Set the bag in a warm place. Note on your calendar to check the seeds for germination in 2 days. After the first inspection, check daily.

5. Plant the sprouted seeds in individual containers or directly in the garden. Handle the seedlings with care so that you don't break the delicate roots and stems.

Guide to Starting Annual Flower Seeds

Most annual flower seeds are easy to start indoors. And by starting your own bedding plants, you'll not only save money, but you'll also be able to grow hundreds of cultivars and unusual species not available at the garden center.

Ageratum (Ageratum houstonianum)

Butterflies love ageratum's fuzzy blue, pink, or white blooms. Grow these mounding plants in beds or containers and be sure to cut some stems for indoor arrangements, too.

To start seeds indoors: Start the seeds 6 to 8 weeks before the last frost date. Scatter the seeds on the surface of the soil—don't cover them because they need light to germinate. Keep the soil temperature at 70° to 80°F; sprouts should appear in 5 to 10 days. Transplant the seedlings outdoors after all danger of frost has passed.

Snapdragon (Antirrhinum *spp.*)

Snapdragon flowers come in all major colors except blue, as well as in combinations of two and three hues. The long flower spikes of the upright cultivars are excellent for cutting gardens; dwarf cultivars make colorful groundcovers.

To start seeds indoors: Sow seeds of upright cultivars 6 to 8 weeks before the last frost date; start dwarf cultivars about 4 weeks before the last frost date. Scatter seeds on the surface of the medium, but don't cover them—they need light to germinate. Keep the soil temperature at 70° to 75°F; sprouts should appear in 10 to 20 days. Transplant the seedlings outdoors after the last frost.

Vinca (Catharanthus roseus)

These bushy 12- to 14-inch plants bear a profusion of single blooms in white, apricot, pink, or rose. 'Stardust Orchid', a 2000 AAS winner, produces big blooms and is easy to grow.

To start seeds indoors: Start the seeds indoors about 8 to 10 weeks before your last frost date. Sprinkle seeds over the moist medium, cover them with a scant ⅛ inch of soil, then lay damp newspaper over the soil surface to keep the seeds moist and dark. Maintain the soil temperature at 70° to 75°F. Check daily for signs of germination; remove the paper as soon as sprouts appear (2 or 3 days). Plant the seedlings outdoors 2 to 4 weeks *after* your last frost date.

Celosia (**Celosia argentea**)

Whether you grow the "plumed" type, which looks like big feathers, or the "crested" type, curled like the combs of a rooster, celosia adds bold color to beds or containers. The stems are good for cutting, too.

To start seeds indoors: Sow seeds 4 to 6 weeks before the last frost date, barely covering them with medium. Keep the soil temperature at about 75°F. When seeds germinate (about 5 to 10 days), lower the temperature to 65° to 70°F. Feed 2-week-old seedlings with diluted fish emulsion and repeat 2 weeks later. Set out the seedlings 2 weeks *after* the last frost date; transplanting them sooner can hurt flowering later in the season.

Cosmos (**Cosmos** *spp.*)

Cosmos bears single white, pink, lavender, or burgundy blooms from summer through frost. Dwarf cultivars, such as 'Sonata', tend to bloom earlier than taller-growing types.

To start seeds indoors: About 4 to 6 weeks before the last frost date, sow the seeds, covering them with about ⅛ inch of medium. Keep the temperature at about 70°F. When seeds sprout (5 to 7 days), lower the heat to about 60°F and provide bright light to keep them from becoming "leggy." Transplant seedlings outdoors around the last frost date.

Globe Amaranth (**Gomphrena** *spp.*)

The upright, branching plants of globe amaranth bear cloverlike red, purple, pink, and white blooms that are excellent for drying. Pick the flowers often to keep the blooms coming.

To start seeds indoors: Start seeds indoors 6 to 8 weeks before the last frost date. Barely cover the seeds with soil and keep the temperature at about 72°F. Germination should occur in 10 to 14 days. Transplant seedlings outdoors around the last frost date.

Sunflower (**Helianthus annuus**)

Popular sunflowers have finally made their way out of vegetable gardens and into flowerbeds, entryways, and even

Here's How Much You'll Save

If you've been buying bedding plants at the garden center each spring, you can save a lot of money by starting your own plants from seed. Here's a comparison of prices for seeds from a major mail-order supplier and the cost of buying an equal number of plants from a garden center (based on $11.99 for a flat of 36 plants).

Flower Packet Price (# of seeds/packet)	Cost at Garden Center	Savings to You
Begonia $5.38 (1,000)	$333.06	$327.68
Celosia $1.75 (150)	$49.96	$48.21
Cosmos $1.75 (40)	$13.32	$11.57
Gomphrena $1.15 (50)	$16.65	$15.50
Impatiens $3.15 (50)	$16.65	$13.50
Marigold $1.70 (20)	$6.66	$4.96
Melampodium $1.65 (20)	$6.66	$5.01
Petunia $2.15 (100)	$33.30	$31.16
Vinca $1.35 (50)	$16.65	$15.30
Zinnia $1.00 (100)	$33.30	$32.30
Grand Total Savings		**$505.17**

containers. Experiment with some of the newer cultivars that offer bi-colored blooms, bushy shapes, and dwarf size.

To start seeds indoors: Sow seeds 4 to 6 weeks before the last frost date. Cover the seeds with ½ inch of soil and keep the temperature at 70° to 85°F. Seedlings will emerge in 5 to 14 days. Set out the plants after the last frost.

Impatiens (Impatiens wallerana)

Arguably the best flower for shady areas, impatiens bloom nonstop from the time you transplant them to the garden until frost. The 6- to 30-inch tall plants cover themselves with single or double blooms in pink, red, purple, orange, salmon, and white—both solid shades and bicolors.

To start seeds indoors: Sprinkle the seeds on the surface of the premoistened starting medium, about 8 to 10 weeks before your last frost date. Don't cover the seeds—they need light to germinate. Set the containers directly beneath fluorescent lights; keep the soil moist and at 70° to 80°F. Seeds should sprout in 1 to 3 weeks. Transplant the seedlings outside a week or two after the last frost date.

Alyssum (Lobularia maritima)

Alyssum's tiny, fragrant flowers of white, pink, or purple attract beneficial insects and butterflies. Use these ground-hugging plants to edge beds, borders, or containers.

To start seeds indoors: Sow the seeds on the surface of the medium without covering them, about 6 to 8 weeks before the last frost date. Keep the soil temperature at 60° to 75°F. Germination should occur in 1 or 2 weeks. Transplant the seedlings outdoors around the last frost date.

Medallion Daisy (Melampodium paludosum)

Melampodium bears beautiful small, golden, daisylike blooms on bushy plants that tolerate heat, drought, and pests. Grow them as an edging plant or in containers.

To start seeds indoors: About 14 weeks before your last frost date, sprinkle the seeds over a premoistened medium, then cover the seeds lightly with vermiculite. Keep the temperature at 65° to 70°F. Sprouts should appear in a week or two. Transplant the seedlings to the garden 2 weeks *after* your last frost date.

Petunia (Petunia × hybrida)

Starting petunias from seed opens the door to a wide world of beautiful, tubular flowers in shades ranging from striking magenta to subdued lavender to pure white.

To start seeds indoors: Sprinkle seeds onto the surface of the pre-moistened medium, 10 to 12 weeks before the last expected frost. Don't cover the seeds—they need light to germinate. Set the flat beneath lights and keep the soil temperature at 70° to 80°F. When seedlings appear (about 10 days), lower the temperature to about 60°F. Transplant seedlings outdoors around the last frost date.

Marigold *(Tagetes spp.)*

Marigolds thrive in just about any climate. And although you can sow the seeds directly in your beds in spring, you'll be able to enjoy their sunny yellow, orange, and bicolored flowers much earlier if you start the seeds indoors.

To start seeds indoors: Sow the seeds about 6 weeks before your last frost date, covering them with a very thin layer of soil. Keep the soil temperature at 70° to 75°F. Sprouts will appear in 5 to 7 days. Plant the seedlings outdoors anytime after the last frost.

Zinnia *(Zinnia spp.)*

Zinnias offer an assortment of bloom colors and shapes on plants that range from 1 to 4 feet tall. Many are great for cutting. If you garden in a humid area, try the newer mildew-resistant varieties, such as 'Oklahoma' or 'Profusion'.

To start seeds indoors: Start *Z. elegans* cultivars 4 weeks before your last frost date; start slower-growing *Z. angustifolia* 6 to 8 weeks before that date. Sow the seeds in individual peat pots, barely covering them with soil. Maintain a temperature of 70° to 80°F until the seeds sprout—about a week. Then, lower the temperature to about 60°F. Transplant to the garden around the last frost date.

Soften Up Tough Seeds

Lupines, baptisia, morning glory, and sweet peas are among the flowers that germinate more easily if their hard seed coats are *scarified*. In this technique, the seed coat is nicked so that the seed will absorb water more readily. Use a knife or fingernail file to nick larger seeds; rub medium-size or small seeds between two sheets of sandpaper. Be careful not to damage the embryo inside the seed.

If you need to scarify a large quantity of seeds, slip a sheet of coarse-grit sandpaper inside a jar. Add the seeds, screw on the lid, then shake the jar until the seed coats wear down.

After scarifying seeds, soak them in lukewarm water for several hours before planting.

A sandpaper-lined jar works well for nicking the seed coat of many seeds at a time.

Starting Avocados from Seed

Want to start an avocado plant from a pit? Forget about the common advice of suspending the pit (or seed) in a glass of water using toothpicks. That method dries out the pit, and transplanting the seedling to a pot is tricky.

Instead, simply bury the pit (pointed end up) in a pot of soil, water it, and leave it alone. After the pit sprouts, the avocado will quickly grow into a large-leaved, graceful small tree that makes an attractive houseplant. (Avocados are hardy outdoors only in Zones 10 and warmer.)

Vinh Pham
Stockton, California

Guide to Starting Perennial Flower Seeds

Growing perennials from seed is the least expensive and easiest way to produce a large number of plants for your landscape. The trade-off is time: You'll need to wait an extra season or two for plants to reach blooming size.

Columbine (Aquilegia *spp.*)

In spring or early summer, graceful columbines produce starlike blooms atop slender, upright stalks. Because the flowers self-sow, plantings tend to increase over the years. Zones 3 to 10.

To start seeds indoors: Sow seeds on the surface of the medium about 8 weeks before the last frost date. (If the columbine seed is not new, refrigerate it for 2 or 3 weeks before sowing.) Keep soil moist and at 70° to 75°F. After the seeds germinate (3 or 4 weeks), lower the temperature to below 70°F. Transplant seedlings to the garden around the last frost date.

Coreopsis (Coreopsis *spp.*)

Sporting long-lasting golden yellow flowers that thrive in full sun, coreopsis often blooms its first year from seed. Zones 3 to 10.

To start seeds indoors: Sow seeds on the surface of the medium 8 weeks before the last expected frost. Keep soil moist and at 65° to 70°F. After seedlings emerge (9 to 12 days), lower the temperature to below 65°F. Transplant seedlings to the garden around the last frost date.

Dianthus (Dianthus *spp.*)

Carnations, pinks, sweet Williams—they're all members of this spring- and summer-flowering genus. The plants produce fringed flowers in shades of pink, red, white, and yellow their first year from seed. Zones 3 to 10.

To start seeds indoors: Sow seeds 4 to 8 weeks before the last frost date. Cover the seeds with ⅛ inch of medium, and keep the medium moist and between 60° and 70°F. Seedlings will emerge in 2 or 3 weeks. Transplant them to the garden around the last frost date.

Purple Coneflower (Echinacea *spp.*)

This native North American wildflower lights up gardens from early summer to frost with purplish pink daisylike blooms. Bees and butterflies like it, too. Zones 3 to 10.

To start seeds indoors: About 12 weeks before the last expected frost, stratify the seeds by refrigerating them for 4 weeks. (See "Give

Seeds the Cold Treatment" on this page.) Then sow the seeds in a fine, moist medium, barely covering them. Keep the temperature at 70° to 75°F. Germination will occur 10 to 21 days later. Plant the seedlings in the garden around the last frost date.

Blanketflower (**Gaillardia**)

These native North American plants bear crimson, yellow, or bi-colored blooms from early summer through fall. Zones 3 to 8.

To start seeds indoors: Sow seeds on the surface of the medium about 8 weeks before the last expected frost. Keep the medium moist and at about 70°F until seedlings appear, then lower the temperature to about 60°F. Seedlings will emerge in 15 to 20 days; transplant them to the garden around the last frost date.

Coral Bells (**Heuchera** *spp.*)

The dainty red, pink, or white flowers of this native North American plant appear in spring or summer, depending on the variety. The handsome foliage is evergreen except in the most severe winter climates. Zones 3 to 10.

To start the seeds indoors: Sow seeds on the surface of the medium 8 weeks before the last average frost date. Keep the medium moist and at 65° to 70°F. When seedlings appear (2 or 3 weeks), lower the temperature to about 60°F. Transplant seedlings to the garden around the last frost date.

Shasta Daisy (**Leucanthemum × superbum**)

The cheerful-looking, single white blooms of Shasta daisies, a favorite for perennial gardens, appear in midsummer. Zones 5 to 8.

To start seeds indoors: Sow the seeds about 8 weeks before the last expected frost. Cover the seeds with ⅛ inch of soil and keep the flats at 60° to 70°F. The seedlings should emerge in 10 to 18 days. Transplant them to the garden around the last frost date.

Rudbeckia (**Rudbeckia** *spp.*)

These easy-to-start perennials are North American natives. They have yellow gold (and/or crimson) daisylike petals with dark centers. Flowering begins in early summer and lasts until frost. Zones 3 to 10.

To start seeds indoors: Sow seeds on the surface of moist soil about 8 weeks before the last frost. (Exception: Refrigerate the seeds of *Rudbeckia fulgida*, black-eyed Susan, for 2 weeks before sowing.) Keep the soil at 70° to 80°F—more seeds will sprout at the higher temperature. Seedlings will emerge in 7 to 21 days. Transplant them to the garden at about the last frost date.

Give Seeds the Cold Treatment

Some seeds—such as those of purple cone-flower, wild bleeding heart, and cardinal flower—must be exposed to cold before they will break dormancy. *"Stratification"* is a technique that simulates natural weather conditions when seeds overwinter in cold, moist ground.

To stratify large seeds, layer them between damp sphagnum moss in a covered container. Mix smaller seeds with damp sphagnum, peat moss, or vermiculite, place them in a plastic bag, and close with a twist-tie. Label the seeds with name and date. Refrigerate for 1 to 4 months, or as directed for the particular cultivar, then sow them in flats or pots.

Germinating and Transplanting Key

	FOR POPULAR FLOWERS				
FLOWER	WEEKS BEFORE LAST FROST TO START SEEDS INDOORS	BEST SOIL TEMPERATURE (°F) FOR GERMINATION	SEED PLANTING DEPTH (IN INCHES)	AVERAGE DAYS TO GERMINATION	COMMENTS
Annuals					
Ageratum	6–8	70–80	Surface	5–10	Set plants out or direct-seed after last frost.
Alyssum, sweet	6–8	60–75	Surface	8–15	Set plants out around last frost. Or, direct-seed 2–3 weeks before last frost. Often self-sows.
Cleome	4–6	Prechill seeds in refrigerator 1–2 weeks; then 70–85.	Surface	10–14	Set plants out or direct-seed just after last frost. Often self-sows.
Coleus	8–10	70–75	Surface; place pot in plastic bag until seedlings appear.	10–12	Tender perennials grown as annuals. Set plants out after last frost.
Impatiens (I. wallerana)	8–10	70–80	Surface; place pot in plastic bag until seedlings appear.	7–20	Set plants out 1–2 weeks after last frost. Susceptible to soilborne diseases; use sterile equipment.
Marigold	6–8. Start triploids and African marigolds indoors.	70–75	⅛–¼	5–7	Set plants out after last frost. Start French and signet marigolds indoors or direct-seed after last frost.
Nasturtium	4–6. Does not transplant well; use peat pots.	65	¼	7–12	Set plants out after last frost. Or, direct-seed 1–2 weeks before last frost.
Petunia	10–12	70–80	Surface; place pot in plastic bag until seedlings appear.	7–20	Set plants out after last frost.
Snapdragon	6–8	70–75	Surface	10–20	Set plants out after last frost. In cool climates, direct-seed a few weeks before last frost.
Sunflower	4–6	70–85	½	5–14	Set plants out or direct-seed after last frost.
Zinnia	4–6. Does not transplant well; use peat pots.	70–80	¼	5–7	In warm climates, direct-seed after danger of frost has passed
Perennials					
Aster	8–10	Prechill seeds in refrigerator for 2 weeks; then 70–75.	⅛; place pot in plastic bag until seedlings appear.	14–36	Set plants out after last frost. Or, direct-seed in early spring or late fall.
Astilbe	6–8	60–70	Surface	14–28	Set plants out after last frost. Or, direct-seed in early spring or fall.
Bee balm	4–8	60–70	⅛	14–21	Set plants out about 1 week before last frost. Or, direct-seed 2–4 weeks before last frost.
Black-eyed Susan	6–8	70–75. Prechill seeds of R. fulgida 2 weeks before sowing.	Surface	7–21	Set plants out after last frost, or direct-seed 2 weeks before last frost.

	Flower	Weeks Before Last Frost to Start Seeds Indoors	Best Soil Temperature (°F) for Germination	Seed Planting Depth (in Inches)	Average Days to Germination	Comments
Perennials—Continued	Chrysan- themum	6–10 before planting out.	60–70	Surface	7–28	Set plants out after last frost. Or, direct-seed in spring or summer.
	Coreopsis	6–10	65–70	Surface	14–25	Set plants out after last frost. Or, direct-seed in spring or summer.
	Daylily	8–10 before planting out.	Refrigerate in moist growing medium in plastic bag for 6 weeks; then 60–70.	⅛	21–49	Set plants out after last frost. Or, direct-seed in late fall or early spring. Takes 2–3 years from seed to bloom.
	Delphinium	8–10	Prechill seeds in refrigerator for 1 week. Thereafter, many species prefer 65–75.	¼	7-21	Set plants out in after last frost. Or, direct-seed in summer for flowers next year.
	Dianthus	4	60–70	⅛	14–21	Set plants out or direct-seed in early spring.
	Poppy	6–10. Does not transplant well; use peat pots.	55	⅛ (most species); surface sow *P. orientale*.	10–20	Set plants out after last frost. Or, direct-seed in late fall or early spring.
	Yarrow	8–10 before planting out.	60–70	Surface	10–15	Set plants out or direct-seed in early spring.

FOR POPULAR VEGETABLES

Vegetable	Best Soil Temperature (°F) for Germination	Average Days to Germination	Weeks to Transplant Size (from sowing)	Spring Setting-Out Dates (relative to frost-free date) Weeks Before	Weeks After*	Comments
Beans, snap	75–80	7	3–4	—	8	Treat roots gently when transplanting.
Beets	75	7–14	4	4	—	Disturb roots as little as possible.
Broccoli	60–75	5–10	6–8	4	2–3	Transplants well.
Brussels sprouts	68–75	5–10	6–8	4	2–3	Transplants well.
Cabbage	68–75	5–10	6–8	5	2–3	Direct-seed mid- and late-season varieties.
Cauliflower	68–86	5–10	6–8	4	2	Direct-seed fall and winter crops.
Cucumbers	70–86	7–10	4	—	8	Often direct-seeded; transplants shorten time to harvest.
Eggplants	70–86	10	6–8	—	2–3	Transplant well.
Lettuce (all kinds)	68–70	7–10	4–6	2	3	Direct-seed some, too, to stagger harvest.
Melons	80–86	4–10	4	—	2–3	Often direct-seeded; transplants shorten time to harvest.
Onions (from seed)	68–70	10–14	4–6	6	2	Plant onion sets at same time seedlings are set out.
Peppers	75–85	10	6–8	—	2–3	Transplant well.
Squash, summer	70–95	7–10	4	—	4	Often direct-seeded; transplants shorten time to harvest.
Squash, winter	70–95	7–10	4	—	3–4	Often direct-seeded; transplants shorten time to harvest.
Swiss chard	50–85	7–10	4	3–6	—	Disturb roots as little as possible.
Tomatoes	75–80	7–14	6–10	—	4	Can plant earlier with protection.

*Indicates latest commonly observed setting-out date. With adequate protection from heat, insect pests, drought, and such, the dates may be extended.

What to Do When Caring for Seedlings

After seeds have germinated and become seedlings, they need different conditions. Here's what to do (and what not to do, on the opposite page) to get healthy seedlings:

- **Do** remove plastic covers to reduce humidity and increase light.

- **Do** cut down on water— let the top ¼ inch of soil dry between waterings.

- **Do** water carefully, either by pouring water into a tray below or by watering between seedlings—not on top of them.

- **Do** transplant before the roots of adjacent seedlings become tangled together—the best time is when seedlings develop their first set of true leaves.

- **Do** transplant each seedling into a clean container of its own.

- **Do** harden-off seedlings for a week or so before you transplant them to the garden.

Questions and Answers about Starting from Seeds

Q: I'd like to grow open-pollinated vegetable varieties this year so that I can save my own seeds for planting next year. When ordering seeds, how can I tell whether a variety is open-pollinated or a hybrid?

A: Open any seed catalog. If a plant variety is not designated as a hybrid (either in the variety name itself or in the description), it's an open-pollinated variety. Seed packets and containers also must be marked if the contents are hybrid.

Hybrids sometimes are indicated by the term "F1" instead of the word "hybrid." This means the seeds are from the first generation of hybrid plants produced by specific and carefully selected parents.

Q: I sowed seeds a month ago and they haven't sprouted, even though the usual germination time is 1 to 2 weeks. What did I do wrong?

A: If you don't see sprouts a couple of weeks after the normal germination time, the following conditions could be to blame:

- The seeds were too old or improperly stored (too damp or too hot).

- The temperature was much higher or lower than what the seeds prefer for germination.

- You planted the seeds too deep.

- The soil was too dry.

Q: One day my seedlings looked fine; the next day they withered and keeled over. I don't think it was due to lack of water. What happened?

A: There's a good chance you've encountered damping-off, a pesky fungal disease that attacks stems near the soil surface and usually causes a plant's demise. Here's how to prevent it:

1. Always use a high-quality seed-starting mix. Drench the mix with a solution of 1 tablespoon of clove oil and a drop of liquid dishwashing detergent in 1 gallon of water. (Scientists recently discovered that clove oil suppresses damping-off fungi. The oil is available through most pharmacies.) Or put ½ inch of milled *light-colored sphagnum moss* (sold as "no damp-off seed starter") on top of the soil mix and then plant your seeds in the moss.

2. Excess moisture or nitrogen can promote damping-off. It helps to thin crowded seedlings, use a fan to circulate the air, water lightly, and wait to fertilize until seedlings are several inches tall.

3. Provide strong light. Keep seedlings on your brightest windowsill or directly beneath fluorescent lights.

Q: My lettuce seedlings are very leggy because I started them on windowsills rather than beneath fluorescent lights. Is it reasonable to expect them to produce good lettuce or do I need to start over?

A: To grow healthy seedlings on a windowsill, the window must have very bright exposure. Otherwise, you will get tall, spindly seedlings that may survive when you transplant them but won't do as well as stronger seedlings.

You don't need a fancy light setup. Simple two- or four-tube fluorescent light fixtures, sold at hardware stores and home centers as "shop lights," can be hung in your basement or a corner of a spare bedroom. Place young seedlings within 2 or 3 inches of the tubes but not close enough to touch them. For seed starting, standard fluorescent tubes will work as well as more-expensive grow lights.

Q: My seeds germinated just fine, but now that the seedlings are 6 weeks old, the leaves are beginning to turn yellow. What should I do?

A: Although seeds need no added nutrients to germinate, older seedlings have requirements that aren't always met by seed-starting mixes. If your seedlings' leaves are pale or yellow near the tops of the plants, a nitrogen deficiency may be the cause. To correct the problem, transplant the seedlings into a more nutrient-rich medium (such as the organic transplant mix on page 46), and feed them every 10 days to 2 weeks with diluted fish emulsion. Yellowing of lower leaves may indicate a magnesium deficiency, which can be corrected by adding dolomitic limestone to the potting mix.

Q: Two weeks after emerging, my tomato seedlings turn purple, become sickly, and often die. If they live, the plants are small. How can I prevent this next time?

A: The purple color and stunted growth are classic symptoms of a phosphorus deficiency. Next year, add some compost to your seed-starting mix to ensure that the young seedlings receive adequate levels of phosphorus and other nutrients. After the seedlings' true leaves emerge, feed the plants once a week for 3 weeks, using a half-strength solution of fish emulsion. After that, feed plants every 2 weeks, using the standard-strength solution recommended on the product label.

What *Not* to Do When Caring for Seedlings

For healthy seedlings, avoid these practices:

♦ **Don't** handle seedling stems—they're fragile and easy to crush. Instead, dig up each seedling with a houseplant trowel or ice-cream stick, then hold it by the leaves to move it.

♦ **Don't** leave seedlings lying out when transplanting them. Have pots ready before you start, then transplant one seedling at a time.

♦ **Don't** fertilize recently transplanted seedlings or you'll shock them. After they're reestablished, give them a boost with liquid seaweed at one-third or one-half strength.

♦ **Don't** let seedlings overheat. Around 60°F is fine for most species; when in doubt, keep them cool.

Check Seed Viability

Should you take a chance by planting leftover seeds that are several years old? The best way to decide is to do a germination test. Here's how:

Write the name of the cultivar on a paper towel and moisten it with water. Count out at least 10 seeds and place them on the towel. Roll it up and put it in a plastic bag, then put the bag in a warm (70°F) place. Check daily for germination. Count the number that have sprouted and compare with the number of seeds you started with. If the germination rate is low, buy new seeds or plant the old seeds a little more generously to compensate for the lower germination.

Q: Can you suggest a recipe for organic seed-starting mix?

A: A good organic seed-starting mix should always contain a generous dose of well-matured compost to provide a steady supply of slow-releasing nutrients and also protect young seedlings from diseases such as damping-off. Here's a recipe developed at Woods End Research Laboratory in Mount Vernon, Maine: To make about 5 gallons of seed-starting mix, combine 1½ gallons aged compost (at least 8 months old), 3½ gallons peat, 1 quart perlite, and ¾ cup dolomitic limestone. You can boost the mix with a well-balanced fertilizer by blending in 1 tablespoon each of rock phosphate and kelp meal and 3 tablespoons of bloodmeal. Or simply water the seedlings every 10 to 14 days with a weak solution of fish emulsion beginning shortly after their first true leaves appear.

Q: I've had no luck germinating the seeds of exotic hot peppers, such as habanero, pequin, and chiltepin. Que pasa?

A: For best germination, these peppers need to be kept *really* warm. (Unlike common hot pepper varieties, these tropical types haven't yet adapted to temperate climates.) Keep daytime temperatures at about 85°F and night temperatures no less than 65°F. Try putting the seed flat in an oven with the light on for warmth (but the oven off). Cover with plastic wrap to retain moisture. As soon as the seeds sprout, take the flat out of the oven, remove the plastic, and water the seedlings.

Q: When starting long-day onions from seeds, I grow the seedlings under lights. Does the number of hours of light that the seedlings get affect their early growth? If so, how many hours should I keep the lights on?

A: Young onions need bright light to grow sturdy. But don't give them more than 13 hours daily, or they'll form bulbs prematurely. That's why onions grown in northern latitudes are called "long-day onions"; their bulbing is triggered by the North's long days in late spring and early summer. If you supply your onion seedlings with the same amount of light as they'd receive during the actual daylight hours outside, you won't exceed their natural requirements.

Q: I planted seeds of gloriosa daisies and impatiens at the same time and treated them the same way. The daisies emerged in 2 weeks, but very few of the impatiens sprouted. What went wrong?

A: You probably covered both kinds of seeds with seed-starting mix. Impatiens are one of the many kinds of flower seeds that need to be exposed to light to germinate. Don't cover the impatiens seeds at all; just mist the sowing medium lightly several times a day until the seeds crack open.

Top 10 Tips for Storing Seeds

If you have leftover seeds or aren't ready to plant your seeds when they arrive, you'll need to store them properly to ensure good germination.

1. Think dry and cool no matter where you store seed. Humidity and warmth shorten a seed's shelf life.

2. The refrigerator is generally the best place to store seeds.

3. Keep seed packets in plastic food storage bags, plastic film canisters, Mason jars with tight-fitting lids, or glass canisters with gasketed lids.

4. Keep your seed-storage containers well away from the freezer section of your refrigerator.

5. To keep seeds dry, wrap 2 heaping tablespoons of powdered milk in 4 layers of facial tissue, then put the milk packet inside the storage container with the seed packets. Or add a packet of silica gel. Replace every 6 months.

6. Store each year's seeds together and date them. Because most seeds last about 3 years, you'll know at a glance which container of seeds might be past its prime when planting season comes.

7. When you're ready to plant, remove seed containers from the refrigerator and keep them closed until the seeds warm to room temperature. Otherwise, moisture in the air will condense on the seeds, causing them to clump together.

8. If you're gathering and saving seeds from your own plants, spread the seeds on newspaper and let them air dry for about a week. Write seed names on the newspaper so there's no mix-up. Pack the air-dried seeds in small paper packets or envelopes, and label with plant name, date, and other pertinent information. Remember, if you want to save your own seeds, you'll need to plant open-pollinated varieties. They'll come back true; hybrids won't.

9. Or dry saved seeds on paper towels. They'll stick to the towels when dry, so roll them up right in the towel to store them. When you're ready to plant, just tear off bits of the towel, one seed at a time, and plant seed and towel right in the soil.

10. Even if you're organized, methodical, and careful about storing seeds, accept the fact that some seeds just won't germinate the following year. Home gardeners will find that stored sweet corn and parsnip seeds, in particular, have low germination rates, and other seeds will only remain viable for a year or two.

GARDENER TO GARDENER

Seed-Starting Cutworm Collars

My seedling pots do double duty as cutworm collars when I set my transplants out into the garden. To make the pots, buy some ordinary paper cups (the size you'll need for your transplants) from the supermarket. Using scissors or a knife, poke a few drainage holes in the bottom of each cup. Fill the cups with moist seed-starting mix, then sow your seeds.

To convert a cup into cutworm collars at transplanting time, use scissors to cut up one side, starting from the bottom, to about 2 inches from the top. From that point, cut around the cup, leaving only the top intact. Place this ring around vulnerable seedlings.

Kay Goyette
Fairfield, Connecticut

Newspaper Seedling Pots

It's easy to make your own biodegradable seedling pots. Simply spread open a standard sheet of black-and-white newspaper, then lay a 1¼ -inch-diameter dowel along one edge of the paper. Roll the paper and dowel one turn, then dab a small amount of flour-and-water paste on the rolled portion of the paper.

Continue to roll the dowel to within 3 inches of the end of the paper, then apply more paste in a zigzag pattern to this remaining area and finish rolling. Remove the dowel and allow the paper to dry overnight. The next day, when the paper is dry, cut the tube into 3-inch lengths.

When you're ready to start sowing, stand the open-ended cylinders upright inside a planting tray or flat, fill each with seed-starting mix, then plant your seeds. When it's time to transplant, place the pots right in the garden—the paper will decompose. (Be sure to cover the entire paper pot with soil so that the paper doesn't act as a wick, drawing moisture away from the seedling roots.)

George Timblin
Brooks, Oregon

Best-Ever Seed Tape

It's difficult to space tiny seeds like those of carrots in the garden, so I make homemade seed tapes by prespacing the dark seeds over a white surface. Here's how to do it:

1. Unroll a strip of toilet paper on a table, mist it with a sprayer, and place the seeds along the center of the paper. (Alternate carrot seeds with radish seeds because when the radishes sprout, they help to mark the row and break the ground.)

2. Starting along the strip's long edge, fold a third of the paper over the seeds, then fold the other third over. Tamp the paper, misting it again to secure the seeds. Make as many of these strips as you need. Then fold the strips up and carefully carry them to the garden.

3. Make shallow furrows in the prepared soil, lay the strips down, and cover them. In a jiffy, your carrots will be planted and perfectly spaced.

Doris Ekblad
Prairie Farm, Wisconsin

GARDENER TO GARDENER

Cooking with Gas

I tried using my water heater and refrigerator as a source of warmth for starting seeds indoors, but the method wasn't working for me. My fridge was too cool for warm-season plants, such as peppers and eggplants, and the water heater was so hot that it quickly dried out the soil.

Now I germinate the seeds before I plant them in the soil using a gas oven with a pilot light that keeps temperatures between 75° and 80°F. This gives me 80- to 90-percent germination rates within a week for every vegetable and herb I've started.

To germinate the seeds, I spread out a few onto coffee filters, then stack the filters three or four layers deep in a metal or ceramic mug, moisten them, and pop them into the oven. (Make sure you put a note on the oven door to warn cooks!) Keep the filters moist and check them periodically to see if the seeds have germinated. When the roots emerge, use tweezers to carefully lift the seeds from the filter and plant them in containers.

Virginia Matzek
Oakland, California

Big Tomato Transplants without "Potting Up"

I like to start my tomato seeds extra early (January or February instead of the usual March or April) so that I can harvest ripe tomatoes from the garden a month ahead of the usual date. Problem is, starting seeds this early often results in spindly plants and crowded root systems. My solution has been to start with large containers and add soil as the plants grow, instead of repotting several times.

I make planters out of clear 2-liter soda bottles. First, I cut off the top of each bottle, then fill the bottom with 2 inches of potting soil. I plant just a seed or two into each bottle. After the seeds sprout, I thin the plants to a single seedling per container. As the seedlings grow, I gradually add more soil, covering the stem up to the first two "true" leaves. I also wiggle the stems a couple of times each day to make them sturdy.

Howard Fuchs
Racine, Wisconsin

Thrifty Plant Labels

You can get a lifetime supply of plant labels from a single plastic miniblind, which costs only about $5. Using a pair of scissors, just snip the slats into pieces and write your plant names on them with an indelible marker. The slats are thin, so you can easily push them into the soil.

David Roth-Mark
New Harmony, Indiana

Seed-Packet Sales Brochure

I live in a rural area that's away from main roads, so I have to use creative marketing techniques to sell my surplus vegetables. To save time, I turn my empty seed packets into a colorful sales brochure that my nieces and nephews show to their city neighbors who might want to buy fresh veggies.

In summer, when my sister and her children visit me, they come with orders in hand. Later, they return to the city with a full load of freshly harvested vegetables to fill their orders.

Ruth Jacobs
Gallipolis, Ohio

GARDENER TO GARDENER

Intensive Seed Starting

If you find yourself "in the dark" at seed-starting time because you don't have enough bright windows in which to set your flats, try vertical seed starting! I rigged up a space-saving seed-starting area by wiring together three open-sided correspondence bins, one on top of the other. Then, because my windowsills are shallow, I suspended the three-tier rack from an overhead curtain rod so that it hangs in the center of a sunny window. (If your business or office doesn't have any old bins to spare, you can buy them at an office supply store.)

Jacque Rutledge
Talihina, Oklahoma

Baker's Seed Flats

The containers that bakeries and delicatessens use for "sheet cakes" are also great for growing seedlings. The containers usually come with a clear plastic dome that fits snugly onto the aluminum pan bottom. The dome makes a perfect minigreenhouse, creating the high humidity that sprouting seeds love.

To use a pan for starting seeds, poke a few drainage holes in the bottom and fill with a moist seed-starting mix. Plant the seeds, then put on the lid. When the seeds begin to sprout, remove the lid and place the container beneath lights. Transplant the seedlings into pots when they have two true leaves.

Kimberly Heal
Hillsboro, Wisconsin

Labels for Life

It's easy to make seed-starting labels from pint-sized plastic containers for yogurt or other foods. Cut 1-inch-wide vertical strips—you'll get 6 to 8 strips from each container—and make a point at the bottom of each strip so that you can push it into the soil. Use a pencil or erasable ink pen to write the name of the cultivar and the sowing date on each label strip. You can reuse these economical labels year after year.

Arthur Kircher
Saint Cloud, Minnesota

Chapter 3

March

Groundwork: Making Compost and Building Soil

It was more than rich, the smell—more even than an odor. An air, a new air would be in the barn, coming out, a new air of life.

—*Gary Paulsen*

The garden work you'll do this month—building your soil—is absolutely critical to the success of your garden. Even with award-winning cultivars and vigorous transplants, your plants and garden won't reach their full potential unless your soil is healthy.

Healthy soil has always been at the heart of organic gardening: Feed your soil organic matter, and your soil will reward you with healthy, productive food crops and lush, beautiful ornamentals. Pest and disease problems in your garden will decrease, too. And the best news is that your efforts are cumulative: The longer you care for your soil, the greater the rewards.

If you're a new organic gardener, you may be surprised at just how simple it is to care for your soil. In most cases, you don't need lots of special fertilizers or gadgets—all you need is compost. If you don't already have a compost pile going, this month is a great time to start. Two prime compost ingredients are (or soon will be) plentiful: leaves and grass clippings.

If you're already making and using compost and your plants still seem to struggle, add "soil testing" to your "To-Do List" for this month. A soil test may reveal nutrient deficiencies that an organic fertilizer can easily fix.

Gardener's To-Do List—March

**If you don't know what USDA hardiness zone you live in,
check the map on page 230 to find out.**

Zone 3

- ☐ Start cabbage, celery, brussels sprouts, broccoli, parsley, onions, and leeks indoors under lights.
- ☐ Start hardy flowers, including snapdragons and pansies, right away.
- ☐ Late this month, start petunia seeds indoors; sow poppies and sweet peas outdoors in the garden (shovel snow first, if you must!).
- ☐ Toward month's end, get a jump start on salad season by sprouting spinach and lettuce indoors.

Zone 4

- ☐ Start cabbage, broccoli, cauliflower, parsley, and celery beneath lights.
- ☐ Trim the tops of onion and leek seedlings to an inch or so high, to keep them stocky.
- ☐ Late in the month, start seeds of head lettuce, eggplant, tomatoes, and peppers indoors.
- ☐ Sprout snapdragons, pansies, and dianthus indoors, then move the seedlings to a coldframe late this month.
- ☐ Begin to prune apples and pears.
- ☐ As snow disappears, begin covering strawberries with row covers to encourage early blooming.

Zone 5

- ☐ Put seed potatoes in a warm, bright windowsill to encourage them to sprout.
- ☐ Start tomato and pepper seeds indoors.
- ☐ Plant new asparagus and rhubarb beds; fertilize established ones with a blanket of compost.
- ☐ Prepare planting beds for cool-weather crops as soon as the soil is dry enough to work.
- ☐ At month's end, move broccoli, cabbage, and cauliflower transplants outdoors to a coldframe. If a hard freeze threatens, cover the frame with an old blanket.
- ☐ Sow sweet peas, poppies, and wildflowers directly into the garden late in the month.
- ☐ Finish pruning apples and grapes, and plant new fruit trees and bushes.

Zone 6

- ☐ Start tomato, pepper, eggplant, and basil seedlings indoors under lights.
- ☐ Move broccoli, cabbage, and cauliflower seedlings to a coldframe.
- ☐ Take a soil sample and have it tested; make the necessary adjustments, using compost and organic fertilizers.
- ☐ Start planting potatoes, peas, spinach, beets, carrots, and radishes late this month.
- ☐ Plant nasturtiums, alyssum, and other half-hardy annuals indoors.

☐ Take cuttings from geraniums and root them in small pots indoors.

☐ Finish pruning fruit trees, then follow up with a dormant-oil spray.

Zone 7

☐ Plant peas, potatoes, and parsley right away.

☐ Begin a new compost pile—using leftover leaves and kitchen scraps—if you don't already have a batch going.

☐ After the last hard frost, transplant cabbage, broccoli, and onions to the garden. Indoors, start seeds of tomatoes, peppers, eggplant, and basil.

☐ Sow carrots, chard, spinach, radishes, beets, and dill in the garden.

☐ Start harvesting asparagus!

☐ Pull winter weeds from flowerbeds, then direct-seed cosmos, rudbeckias, and nasturtiums.

☐ Plant roses, lilies, and all types of hardy groundcover plants.

☐ Fertilize strawberries with compost when the first blossoms show.

Zone 8

☐ Mulch potatoes with straw as soon as the fat sprouts emerge from the soil.

☐ Thin leafy salad greens to keep them growing fast.

☐ Make a second sowing of carrots, beets, chard, and dill early this month.

☐ After midmonth, sow cucumbers, bush beans, and early sweet corn.

☐ Fill in empty spaces in flowerbeds with direct-seeded cosmos, nasturtiums, and lavatera.

☐ Set out gerbera daisies and other hardy perennials, such as roses, lilies, hardy groundcover plants, and woodland shrubs.

☐ Apply compost or other organic fertilizers to fruit trees and shrubs.

Zone 9

☐ Make your last sowings of carrots, beets, and heat-tolerant leaf lettuce, then start planting warm-weather crops.

☐ Start eggplant seedlings and sweet potato slips indoors.

☐ For quick color, sow cosmos and nasturtium seeds right where you want them to grow.

☐ Sow sunflowers and tithonia where their height will mesh with your garden design.

☐ Transplant salvias, marigolds, zinnias, celosia, gomphrena, and other annuals to flowerbeds and borders.

Zone 10

☐ Set out tomato and pepper transplants; plant basil, squash, corn, beans, cucumbers, and melons.

☐ Set out caladium corms in shady spots, or mix them with coleus in containers.

☐ Before slugs, snails, and pill bugs dig into your strawberries and leafy greens, defeat the pests with beer traps.

☐ Plant kiwi vines and avocado trees, and water new plants every few days to keep the roots from drying out.

The Beauty of Compost

Compost is simply decomposed organic matter, but it's the best gift you can give your garden. Here are just a few of the many benefits of compost:

- **Protects plants from drought.** Compost encourages the formation of soil clusters that soak up water and hold it like a sponge.

- **Improves soil aeration.** Adding compost to soil creates air pockets between soil clusters.

- **Stops erosion.** Soil that contains lots of humus (compost) resists erosion from wind and water.

- **Slowly releases nutrients.** Compost acts as a nutrient storehouse, gradually releasing nutrients to plants throughout the growing season.

- **Prevents disease.** A thin layer of compost spread over the soil's surface will fight plant diseases better than any chemical fungicide.

- **Recycles wastes.** Composting makes use of grass clippings, leaves, and kitchen scraps that otherwise would take up space in a landfill.

5 Steps to Fast Compost

Making compost is probably the single most important thing you can do for your organic garden. The success of your garden depends on the soil, and the health of your soil depends on the compost you give it. And making compost isn't difficult. With very little effort on your part, you can turn throw-away materials into this sweet-smelling, nutrient-rich, no-cost soil conditioner.

There are many approaches to making compost, but the fastest way to get finished compost for this year's garden is to make an active, or "hot," compost pile. By providing a steady supply of water and air to the pile, you'll encourage the microorganisms that drive the composting process to work faster. Here are the five key steps for making compost in about 30 days.

1. Shred and chop. Shred or chop materials as finely as you can before mixing them into the pile. For example, you can chop fallen leaves by running your lawn mower over them. The same strategy applies to kitchen scraps and the like—"the smaller, the better" is the rule for compost ingredients.

2. Mix dry browns and wet greens. The two basic types of ingredients for making compost are those rich in carbon and those rich in nitrogen. Carbon-rich materials, or "dry browns," include leaves, hay, and straw. Nitrogen-rich materials, or "wet greens," include kitchen scraps and grass clippings; these work best when used sparsely and mixed in well so they don't mat down. Your goal is to keep a fair mix of these materials throughout the pile.

Mowing leaves results in small pieces that break down fast.

3. *Strive for size.* Build the pile at least 3 × 3 × 3 (or 4) feet so materials will heat up and decompose quickly. (Don't make the pile too much bigger than that, though, or it will be hard to turn.) Unless you have this critical mass of materials, your compost pile can't really get cooking. Check the pile a couple of days after it is built up—it should be hot in the middle, a sign that your microbial decomposers are working hard.

4. *Add water as needed.* Make sure the pile stays moist, but not too wet. (It should feel like a damp sponge.) You may need to add water occasionally. Or, if you live in a very wet climate, you may need to cover the pile with a tarp to keep it from becoming too soggy.

5. *Keep things moving.* Moving your compost adds air to the mix. You can open up air holes by getting in there with a pitchfork. Even better, shift the entire pile over a few feet, bit by bit, taking care to move what was on the outside to the inside of the new pile, and vice versa. Or consider using a compost tumbler, a container that moves the materials for you when you turn it.

A less labor-intensive way to add air is to run ventilation pipes—perforated PVC pipes, cornstalks, or palm fronds—through the pile. Run one pipe horizontally through the middle of the pile; also insert a pair of pipes vertically about 6 inches on either side of the horizontal pipe. As the pile heats up, withdraw a pipe and reinsert it in another area of the pile. Do this every few days, if possible.

For fast, finished compost, turn the pile at least once a week.

Rx for Compost Problems

Because compost is so integral to organic gardening, the creation of a "perfect" compost pile is a goal to which many gardeners aspire. But what if your results are less than perfect?

Don't throw in the towel! Here are some common compost problems and what to do about them.

Soggy, Slimy Compost

Nothing is worse than cold, slimy compost! How does it get this way? Usually there are three contributing factors: poor aeration, too much moisture, and/or not enough carbon-rich materials in the pile.

A compost pile that is overburdened with materials that mat down when wet—such as grass clippings, spoiled hay, and unshredded leaves—can become so dense that the pile's center receives no air. If you leave the pile uncovered during prolonged rain (and you don't turn it to introduce air into the center), you'll end up with a cold, soggy mess.

Aerobic bacteria—the tiny, friendly, air-loving creatures that make compost cook—can't live in such an oxygen-poor environment. That kind of pile welcomes *anaerobic* bacteria instead. These creatures will eventually make compost of your mess, but they work much more slowly than aerobic bacteria and the compost will be soggy and slimy during the 2 or 3 years it takes to decompose.

But there's an easy solution. If endlessly wet weather is part of the problem, cover the pile with a loose-fitting lid or tarp. You'll also need to turn the pile and fluff it up thoroughly. If you have some nitrogen-rich ingredients (such as shellfish wastes) and fibrous, nonmatting materials (such as shredded corncobs or sawdust), add them to help get things cooking.

Dry, Dusty Compost

This problem is common in areas that receive little summer rainfall. The stack just doesn't get enough moisture to support the bacterial life necessary to fuel the composting process. Luckily, curing dry, dusty compost is as simple as turning on the spigot. Add water until the materials are as wet as a damp sponge.

Put an oscillating sprinkler on top of your dry compost pile and run it for about an hour. This will moisturize the materials better than watering with an open hose—especially if the outer layers consist of materials that tend to shed water when they dry out. After sprinkling, check the center of the pile to make sure it's moist. To moisten the entire pile, you might need to turn it and water the layers as you go.

Turning and watering your pile should help it spring to life fairly quickly. If not, you might have other problems, such as a lack of nitrogen-rich materials. If that's the case, tear the whole thing apart, add some grass clippings or alfalfa meal to get it going, and pile it back up. When the pile starts cooking, don't let it dry out again. As those tiny organisms multiply, they use up a lot of water. You might have to water your compost almost as often as you water your roses during a heat wave.

Bugs!

Pill bugs and sow bugs are small crustaceans (not insects) that live on decaying refuse. If you turn your compost pile and see thousands of tiny gray "armadillos" with seven pairs of legs each, you've discovered a nest of these primitive creatures. (Pill bugs roll up into a ball when threatened and sow bugs don't—otherwise, there isn't much difference between the two.)

These bugs won't harm your compost. In fact, they're actually helping to break it down. But if you don't remove them from the finished mixture before you spread it onto the garden, you might find them snipping off the emerging roots and leaves of your garden plants.

Ants and earwigs also invade compost piles. Like sow bugs and pill bugs, they are essentially harmless to the composting process. Their presence may indicate that the compost is nearly finished. Or, if you still have lots of material that hasn't broken down, these insects can simply mean that your pile is on a slow track to decomposition.

No Room to Compost? Try Trenches

If you don't have room for a compost pile, you may find trench composting the perfect solution.

To trench compost, just dig a series of 8-inch-deep holes in your vegetable garden, one each day, and bury your kitchen scraps in them. If you start along one side of the garden, you'll be fertilizing the nearby vegetables as well as boosting soil texture and fertility for the next growing season.

Pill bug

Pill bugs live on the decaying matter in your compost pile. No need to worry—they're harmless.

Trouble Turning?

For seniors, disabled gardeners, and those who can only garden on weekends, turning a compost pile frequently can be difficult. If you can't turn your compost pile, relax—you don't have to! Here's how to get finished compost in 6 months to 1 year, without ever turning the pile:

To ensure complete breakdown of the materials, shred everything *before* you add it to the pile. Put the materials through a chipper/shredder, run a lawn mower over them, or use a pruning shears to cut up large stems.

If that's not possible, build your pile in layers, alternating "browns" (carbon-rich materials like dry leaves and straw) with "greens" (nitrogen-rich materials like grass clippings and kitchen scraps), mixing them together and adding water as you go.

Even if you can't do any of this, not to worry! As long as you can wait a while, an unmanaged pile will turn to compost eventually.

To get these bugs out of unfinished compost, raise the pile's temperature to above 120°F. You can do that by turning the pile and rebuilding it, watering as you go. If it contains lots of slow-to-decompose materials, such as leaves or straw, mix in a nitrogen source, such as alfalfa meal or grass clippings. It should soon start heating and, when it does, the bugs will depart for a more comfortable place.

But what if your sow-buggy compost is already finished and you want to use it on the garden without endangering small plants? De-bug it! Simply spread the compost in a thin layer over a tarp in direct sunlight and leave it there to dry.

Bad-Smelling Compost

If the pile emits the sharp, nose-twisting stench of ammonia, it probably contains too much nitrogen-rich material and is too wet to allow aerobic bacteria to thrive. If it just "smells rotten" and there are lots of flies hanging around, you've most likely added too many kitchen scraps or canning wastes without chopping them first or mixing them in thoroughly. For both cases, you should remake the pile to bring your stinky compost under control.

If you notice the smell of ammonia, turn the pile and add absorbent materials—such as straw, shredded tree leaves, or sawdust—as you go. Rebuild the pile to a height of 3 feet to get it cooking again.

If kitchen scraps, canning wastes, or large amounts of other mucky stuff are producing offensive odors, you can turn the pile without adding anything. As you turn, break up all that mucky stuff and mix it in well. Remind yourself that you'll avoid this unpleasant task in the future by first finely chopping these materials and then mixing them thoroughly into the heap. If odors persist, add more carbon materials, such as straw or shredded leaves, to the pile.

"Burning"

In cold weather, an active compost pile may look as though it is on fire, but in fact it is "breathing" its warmth into the frosty air—a positive sign that things are working as they should.

But if your compost's carbon-nitrogen ratio is off, the center *can* overheat and dry out, leaving visible white streaks. These streaks are the dead bodies of millions of microorganisms. And if they hadn't been cooked, they still would be busy turning your pile into finished compost!

Sometimes called "fire fanging" or "burning" by old-time gardeners, this phenomenon is quite common in piles that have been overloaded with nitrogen. Regular turning and watering will help you keep your compost pile lively.

Raccoons, Opossums, and Other Critters

If you spot raccoons, opossums, dogs, skunks, rats, or bears at your compost pile, they're probably going after the fresh, edible kitchen scraps you recently buried.

To avoid attracting animals, remember to keep meat scraps and fat out of the pile. Mix other scraps with soil and/or shredded leaves, then bury them deeply into the center of the pile.

If animal scavengers have grown accustomed to a free meal from your compost pile, you could have trouble breaking them of the habit. You'll probably have to build or buy a covered container for your compost. Choose among the many commercial compost bins and tumblers on the market, or build a simple enclosure from untreated lumber.

Plants Growing in the Pile

Sometimes young plants emerge from a pile of finished, or nearly finished, compost. Invariably, a few seeds will survive the composting process, even if the pile is hot. And the temperature, moisture, and fertility of the maturing compost are just about perfect for seed germination and seedling growth.

If the sprouts are weeds, just pull them out and add them to your next pile. (Green plants of any kind will provide nitrogen.) But if the young plants are vegetable or flower seedlings, go ahead and transplant them. Just be aware that they may be the offspring of hybrid parents and could, therefore, grow up to be something other than what you grew last season.

Compost Checklist

Use the following checklist to gauge the success of your finished compost:

Structure: Should be loose and crumbly—not too tight (like soil) and not lumpy

Color: Should be black-brown—not pure black or grayish yellow (signs of excess water)

Odor: Should be earthy sweet—not bad-smelling (a sign that decomposition hasn't finished) or musty (indicating the presence of molds)

Moisture: Should be about as moist as a damp sponge

In cold weather, vapors indicate that your compost pile is working!

Top **10** Soil-Turning Tools

Loosening your garden's soil can be a tiresome chore. Make it easier by using any one of these top-quality tools to help lighten the workload.

1. **Broadfork:** Sometimes called a U-bar digger, this tool has five 10-inch-long tines attached to a U-shaped bar. The tines loosen the soil as you push the tool into the earth and pull it out.

2. **Hoes:** Hoes come in dozens of shapes and sizes and serve many functions, from preparing and furrowing soil to weeding and cultivating it. Choose from the oscillating hoe, the warren hoe, or the collinear hoe, just to name a few.

3. **Pickax:** This tool is a necessity when your soil is rocky or full of tree roots. Use the broad hoelike blade to pulverize small rocks and soil clods.

4. **Rakes:** You'll probably want one of each type—a steel garden rake and a steel or bamboo leaf rake. Garden rakes make short work of leveling out ground and creating raised beds, and leaf rakes are good for spreading lightweight mulches and smoothing the finely prepared soil on top of a seedbed.

5. **Shovel:** A must-have soil-care tool with a rounded edge. Handy for scooping compost, cutting into hard ground, and digging soil out of planting holes.

6. **Spade:** A squared-off blade makes cutting through sod a snap. Also useful for digging new beds or borders and for edging.

7. **Spading fork:** Though specifically designed to cut into the soil with its four flat or slightly rounded tines, the spading fork can mix materials into the soil and harvest root crops as well.

8. **Tillers:** Rotary tillers are unsurpassed for breaking new ground, breaking up large soil clumps, digging furrows, and mixing in soil amendments, compost, and cover crops. (But be careful not to overtill because it destroys soil structure.)

9. **Trowel:** This miniature shovel is perfect for digging holes for smaller plants and bulbs and for removing unwanted weeds.

10. **Your hands:** Undoubtedly the most important tools you'll use when working the soil. Protect your hands when using many of the tools listed above by wearing gardening gloves.

Buying Quality Compost

Homemade compost is the best thing you can use to feed your plants and improve your soil. But even if you make your own, there may be times when you don't have enough of it to go around. That's when commercial compost comes in handy. Be forewarned, though: Not all commercial compost is created equal. Commercial composts are highly variable in organic matter and nutrient content. Some can even harm your soil and plants.

Fortunately, a simple look and smell are all you need to do to find a good-quality product. If possible, ask your garden center or supplier to let you take home a few samples before you buy. Put the samples through your own round of testing, using the following criteria:

The texture should be loose and granular. There should be little or no recognizable bark or wood. If the compost isn't loose enough for you to spread and work it easily into your garden beds, don't buy it.

The color should be dark brown. Compost that's light in color probably contains little organic matter and too much soil. It's easiest to tell the true product color if you let the compost sample dry out.

The compost should be moist, not dry or soggy. In the soil, compost can hold up to two times its weight in water. But in bagged products, excess water makes the compost difficult to spread. (Plus, you'll be paying for water, not compost.) Simply lifting a bag of compost will give you a good idea of its moisture content. If it feels like a big glob, the compost probably is too wet. If it feels loose, it's probably drier.

Mature compost has a pleasant, earthy aroma. If you smell an earthy, woodsy odor, you've probably struck "black gold"—a mature, good-quality compost. Avoid composts with a strong or unpleasant smell, indicating immature compost that could damage plants. (*Note:* Some good-quality bagged composts have a slight musty or barnyard odor when first opened. That's because the plastic bags restrict the oxygen supply to the organisms that release the earthy odor. After a day or two, the compost should acquire that earthy aroma.)

Mature compost contains 30 to 60 percent organic matter. To test the organic-matter content of any compost, commercial or homemade, spread some out on a thin layer of newspaper and let it air dry for about a week. Then measure exactly 1 cup of the dried compost and weigh it. If it weighs between 4 and 6 ounces, it contains the desired amount of organic matter. If it weighs less than 4 ounces, it's probably immature. If it weighs more than 6 ounces, it's probably old or diluted with soil, so you shouldn't use it.

Look Before You Buy

Some stores won't allow you to sample or inspect before you buy. If you buy a bag and find that it's not up to these standards, take it back and ask for a refund or dump it onto your home pile to dilute and fully compost it.

Another option is to look for bulk compost, sold at some garden centers. You'll not only be able to see what you're buying, but you'll also save some money. Many bulk composts are cheaper than bagged products: 1 cubic yard of bulk compost (the equivalent of twenty-five 40-pound bags) usually costs less than $30, while the good-quality bagged composts sell for $2 to $4 per 40-pound bag. Another advantage of bulk compost is that the garden center might be willing to deliver it right to your garden.

Guide to Organic Fertilizers

Sometimes even organic gardeners need to take extra steps to improve their soils. Maybe you're new to gardening. Or maybe you garden in a region that's naturally deficient in some nutrient. Or maybe you've been gardening organically for years but have unwittingly caused a soil imbalance by adding too much of a certain amendment. In most cases, you can help get your soil on track by adding the right amount of the right organic fertilizer.

Organic gardeners use fertilizers in the same manner a cook uses herbs and spices: They add the extra touch that brings out the very best in plants. Organically managed soil is biologically active and rich in nutrients. Organic gardeners don't need to pour on chemical fertilizers to get plants to grow well because organic materials serve as both fertilizers and soil conditioners, meaning they feed both soil and plants.

But how do you choose *which* fertilizer—and how much of it should you use? The first step is to have your soil tested. (See "Get Your Soil Tested!" on this page.) Most soil test reports will tell you how many pounds of which nutrients you need to add to correct any problems.

Once you know your soil's specific needs, you can get down to choosing fertilizers. Here are some basics to help you choose the ones that make the most sense for your garden. (For advice on determining exactly *how much* of each fertilizer to use, see page 65.)

Nitrogen Fertilizers

If your soil turns out to be very low in organic matter (an indicator of how much nitrogen is available to plants), you probably need to add some nitrogen to your soil. Nitrogen promotes leafy growth, so crops such as lettuce, cabbage, spinach, and corn all need good doses of it. But don't overdo the nitrogen on other plants, or they'll produce lots of leaves and stems and fewer flowers and fruits. Plants respond quickly to nitrogen fertilizers, so you should err on the side of caution and add more later if necessary.

Bloodmeal. The name's not pretty but it is accurate. This fertilizer is dried and ground blood, and it's a very potent nitrogen source, so use it sparingly. If your soil's organic matter level is low, work in up to 3 to 5 pounds of bloodmeal per 100 square feet. If nitrogen-hungry crops seem to need a boost during the season, side-dress them *lightly* with bloodmeal, watering it into the soil after you apply it. A single application of bloodmeal will last about 3 or 4 months.

Fish meal. Consider giving the garden a soil-enriching treat of dried and ground fish scraps. Compared to bloodmeal, fish meal is usually less expensive, contains more phosphorus, and also contains many trace minerals that plants need. It also breaks down more slowly, lasting about 6 to 8 months in the soil. To correct soil with low amounts of organic matter, work in up to 3 pounds of fish meal per 100 square feet before planting. You also can put a little fish meal into the planting holes of leafy crops and brassicas (but *not* fruiting crops such as tomatoes or peppers) when transplanting them in spring.

Phosphorus Fertilizers

Plants need phosphorus to grow strong root systems, which are vital to their overall health. Plants also need phosphorus to flower well. But phosphorus doesn't move around easily in the soil. You need to dig it in where plant roots can reach it—no deeper than the top 6 to 8 inches of the soil. Here's what to use:

Bonemeal. Bonemeal is exactly what it sounds like—ground animal bones. Most bonemeals are steamed, rather than raw, and say so on the bag. Steamed bonemeal has the advantage of supplying phosphorus faster than the raw product, and so it is more convenient for immediate phosphorus needs. Bonemeal lasts anywhere from 6 to 12 months in the soil, and you should apply up to 3 pounds per 100 square feet if your soil tests low in phosphorus.

Rock phosphate. Rock phosphates are the mined skeletal remains of prehistoric animals, and they release their phosphorus very slowly, over about 5 years. For rock phosphate to break down and be useful to plants, your soil needs to have a lot of worms and microbial life, and its pH should be 6.4 or slightly less. If your soil meets these conditions and tests low in phosphorus, apply up to 5 or 6 pounds of rock phosphate per 100 square feet.

Colloidal rock phosphate. Colloidal (soft) rock phosphate is a byproduct of mining rock phosphate. Compared to regular (hard) rock phosphate, it requires less activity by worms and microbes to make the phosphorus and trace minerals available. Colloidal rock phosphate also contains clay, which helps bind sandy soils together. (If you have clay soil, use regular rock phosphate instead.) If your soil is sandy and low in phosphorus, apply up to 5 or 6 pounds of colloidal rock phosphate per 100 square feet. A word of caution: The particles of colloidal rock phosphate are very fine, so be sure to wear a dust mask when working with it.

Potassium Fertilizers

Potassium helps plants function smoothly by promoting the flow of nutrients through the entire plant. It also improves the quality of fruits and seeds and helps plants withstand stresses, such as disease, drought, and extreme temperatures. If your soil test report indicates a need for potassium (also called "potash"), you have two fertilizer options besides compost, which usually is an excellent source of potassium, too.

Greensand. Also known as glauconite, greensand comes from a 70-to-80-million-year-old marine deposit mined in New Jersey—the only place it occurs. Greensand releases potassium very slowly over a 10-year period. It is also very rich in trace minerals, and some experts feel that these alone make greensand a worthwhile investment. If your soil is low in potassium, apply up to 5 pounds of greensand per 100 square feet of soil in the fall. Organic matter speeds the release of the potassium, so some gardeners prefer to add greensand to their compost, then apply the compost to their soil.

Sul-Po-Mag. It may not sound organic, but Sul-Po-Mag is the commercial name for the mined mineral otherwise known as sulfate of potash-magnesia, and it *is* organic. If your soil needs an immediate potassium boost, this is the fertilizer for you. But be careful not to overuse Sul-Po-Mag though—it can interfere with plants' absorption of other nutrients, such as nitrogen. It also has a high level of magnesium, so try to avoid using it in combination with dolomitic limestone, which also contains magnesium. On low-potassium soils, broadcast up to 1 pound per 100 square feet, then work it into the soil.

Calcium Fertilizers

Calcium is needed in greater quantity than any other soil nutrient. And, especially in the East, it's a nutrient that's often lacking because rainwater leaches it out of the soil. A calcium deficiency can cause dieback of growing tips of plants and roots, and blossom-end rot on tomatoes and peppers.

The best way to correct a calcium deficiency depends on your soil's pH. If it is lower than 6.5 (on the acid side), use limestone. If your soil's pH is 6.5 or higher (neutral to alkaline), use gypsum.

Limestone. Two main kinds of limestone are used in agriculture, *dolomitic* and *calcitic*. (Don't use "hydrated" or "slaked" limestone, or "builder's lime"; these are caustic to earthworms and plants.) Dolomitic limestone contains magnesium (8 percent) as well as calcium (25 percent) and should be used only if your soil test report says that you should add magnesium. Calcitic limestone, sometimes sold as "oyster shell lime," contains little or no magnesium. Both types of limestone will boost the calcium levels in your soil gradually and will slowly raise the pH at the same time. Apply limestone with a garden spreader, adding 2 to 8 pounds per 100 square feet of soil.

Gypsum. Gypsum is a mined calcium sulfate powder that's gray in color; when added to soil it will not affect the soil's pH as limestone will.

Spread up to 4 pounds of granular gypsum per 100 square feet on alkaline soils that test low in calcium. In addition to helping balance your soil's nutrients, gypsum is an excellent soil conditioner; you can use it to loosen clay soil or to bind up sandy soil.

Sulfur Fertilizers

Sulfur works to make nitrogen available to plants; without enough sulfur, plant growth slows and leaves can turn yellow. Soils low in organic matter and with alkaline pH levels are likely to be low in sulfur.

Elemental sulfur. Also known as "soil sulfur" or "agricultural sulfur," elemental sulfur is a naturally occurring mined mineral. It works very quickly to correct sulfur deficiencies and lowers soil pH faster than any other amendment you can apply. But be very careful not to add too much to your soil. For every *full point* that you want to lower your soil pH, spread 1 pound of sulfur per 100 square feet and mix it into the top 3 inches of soil.

Alternatives. If you don't want to add elemental sulfur to your soil, two milder sources of sulfur are Sul-Po-Mag and gypsum. Both contain about 19 percent sulfur, enough to lower soil pH and release nutrients for your plants to use.

How Much to Use

Figuring out exactly how much of a given nutrient to use on your garden can seem tricky, at best. Here's a technique you can use with any fertilizer. Let's say your soil test report advises you to add 2 pounds of phosphorus per 1,000 square feet of soil and you choose bonemeal as your phosphorus source. Bonemeal has a phosphorus content of about 20 percent (it should say so on the label). Divide the number of pounds of nutrient you need (2 pounds) by the percentage of nutrients in the fertilizer (20 percent, or 0.20). That gives you the number of pounds of fertilizer you need (10 pounds, in this case) to add per 1,000 square feet of garden space.

Micronutrients: Who Needs 'Em?

Sometimes called "trace elements," micronutrients are essential to plant health, but as their name suggests, plants use them in very small amounts. They include zinc, iron, and copper, and the best sources are compost and other organic materials. If you haven't been building up your soil and your garden has visible problems, a micronutrient deficiency could be the cause. Ask that your soil be tested for micronutrients when you send in a sample. You'll probably pay more than for a standard soil test, but the resulting information will tell you just what your garden needs in terms of amendments.

Materials

2 untreated 1" × 12" × 8' boards

2-inch galvanized deck screws, or 2-inch galvanized or aluminum nails

¼-inch or ½-inch metal hardware cloth

Heavy-duty staple gun and staples

Technique

Build a Super Compost and Soil Sifter

For starting seeds and filling pots, you need "screened compost." That's where a compost sifter like this one comes in handy. The sifter is a screening box that is designed to fit right on top of your wheelbarrow or garden cart, so that you can fill it with fine-textured compost and then toss the larger remnants back into the compost pile for further cooking. You can construct this useful gadget in just an hour for a cost of less than $15.

1. Measure the inside distance between the sides of your wheelbarrow or cart—this measurement will determine the size of your sifter's opening.

2. From each board, cut one 24-inch-long piece and one 36-inch-long piece. From each of the 26-inch leftovers, cut one ¼-inch-wide retaining strip to help hold the screen in place. (Use a jigsaw to cut handholds, as shown in the illustration. Use a router or sandpaper to smooth the edges of the holds.)

3. Screw or nail the four sides together at the corners.

4. Using the staple gun, staple the hardware cloth to the bottom edges. (For potting soil, use ¼-inch hardware cloth; for other garden uses, ½-inch cloth is fine.)

5. For the bottom of the sifter, cut two 24-inch-long filler pieces from the leftover boards to cover the parts of the box that will extend beyond the sides of your wheelbarrow or cart. (Use the measurement from Step 1 to determine the size of the pieces.) Also cut ¾-inch-wide guide strips from the leftover pieces to hold the sifter tightly in place on your cart. Screw or nail the filler boards into place and then staple the screen to them. Now screw or nail the ¾-inch pieces; attach the retaining strips to the long edges of the box; and attach the guide strips to the edges of the filler boards so they will rest against the facing sides of your wheelbarrow or cart and hold the sifter in place.

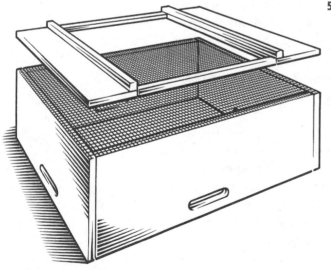

The sifter, shown upside down, is designed to fit your wheelbarrow or cart.

GARDENER TO GARDENER

Compost in 14 Days—Guaranteed

We've been making a finished batch of compost every 14 days for more than a year, with the help of a chipper/shredder, a barrel composter, and a compost thermometer. The compost ingredients consist of grass clippings, dry leaves, organic household waste, chipped branches, spent garden plants, and horse manure. By following our recipe, you too can have finished compost in just 14 days.

Here's what we do. We place one 5-gallon bucket each of aged horse manure, dry shredded leaves, shredded household scraps, and dry grass in the compost drum. Using a hose, we moisten the mixture until it's about as damp as a wrung-out sponge. Then we add another 5-gallon bucket of each of the materials, moisten them, and continue the layering until the drum is almost full. (Because we live in a very dry region, it's important for us to water the ingredients as we assemble them.) When we've finished adding the materials, we wet the whole mixture again. Then, we set up three 5-gallon buckets beneath the compost drum to catch the water that drips out, and put the compost thermometer inside the center support axle of the drum.

Within 24 hours, the compost thermometer reads 160°F, and it remains there for at least 3 days—ensuring that weed seeds and pathogens have been killed. We turn the barrel several times a day and check the moisture content often. If the materials dry out, we empty the buckets of drippings back into the drum.

Jim and Jenni Muir
Peoria, Arizona

Mow Better Beds

Here's a way to lay out a new flowerbed so that it fits into your landscape and allows you to make any necessary revisions before you do any work.

In early spring, when your grass is a bit longer than usual, mow just the area where you plan to put the new bed. Then stand back and take a good look at the effect. If you like what you see, stake off the area, dig up the sod, and prepare your new bed!

Becky Wethington
Kinston, North Carolina

Warming Up Cold Soil

In many regions of the country, cold soil is a fact of life. Fortunately, we've discovered how to warm up our soil so that we can grow a year's worth of vegetables in just 3 months.

To take full advantage of the available warmth, we garden on a south-facing slope. In April, we begin shoveling snow off the site to speed the thawing of the frozen soil. Then, as soon as the soil dries, we till the area to expose the cold subsurface soil so that the sun can warm it.

About 2 weeks before planting cold-sensitive crops (such as corn, squash, melons, and beans) in raised beds, we heat the soil by covering the beds with clear plastic. To plant, we remove the plastic or cut ×s into it. We avoid using other mulches (such as straw) until the soil is sufficiently warm in June, because they do the opposite of what we want—they insulate the cold soil and keep it from warming up!

Miki Collins
Lake Minchumina, Alaska

GARDENER TO GARDENER

Compost from Sun and Sea

I need large quantities of compost for my garden, so I devised a system that gives me a wheelbarrow-load of finished compost every week.

My system begins with my homemade composters: four 55-gallon plastic drums (previously used to ship olives) mounted individually on posts. To let air inside and keep rodents and flies out, I cut out a piece from each drum lid and stapled a piece of strong screen to the inside of the lid. Metal clamps hold the lids tightly on the drums. For maximum solar heating, I position the drums so that the sun hits their sides at noon.

Every week, I fill one drum with organic kitchen waste, 3 gallons of dry chicken manure, 10 gallons of seaweed (rinsed with water several times to remove excess salt), and some dry sheep manure.

Initially, I mix the compost by turning the drum several times. After that, I turn each drum once in the morning and once in the evening. If it rains, I turn the drums so their tops are facing the ground to keep out water.

During the first 10 days or so, the compost heats up and begins to steam. The process slows over the next couple of weeks and finishes in about 4 weeks. By filling the four drums on a rotating basis, I have a near-constant supply of ready-to-use compost.

> *Jerry Rudy*
> *Jenner, California*

New Soil—Fast!

When we moved to our new home, we knew we had our work cut out for us. The thin layer of topsoil was mostly sand. To build the soil, I started by turning over the existing sod with a shovel. Then I covered the clumps with 3 or 4 inches of aged grass clippings and soaked the mound of sod and clippings with water. Next, I added two more layers—3 inches of purchased black peat soil and another layer of aged grass clippings—and again soaked the area with plenty of water. To top it all off, I added several inches of quality topsoil from a local supplier. My gamble paid off—I harvested broccoli, beans, and tomatoes well ahead of the usual time in our area.

> *Walter Andrews*
> *Salmon Arm, British Columbia*

The Compost Path

Garden paths are an ideal way to compost twigs and other rough garden debris.

Simply lay twigs, small branches, and grass clippings directly on the paths between raised beds to create a soft, comfortable walkway. Excess rain and irrigation water will drain out of beds and into the materials in the paths. After a month or so, turn the compost and add another thin layer of grass clippings. By fall, you'll have finished compost to turn into beds for the next growing season.

> *Meilie Moy-Hodnett*
> *Rockville, Maryland*

The Milky Way

Organic gardeners know that compost is the best possible amendment for their soil. But when you want to give your vegetables and flowers an extra boost of nutrients, try this: Sprinkle about ½ cup of powdered milk around the base of each plant. Be careful not to get any milk on the plants' leaves. My plants seemed to love it!

> *Paula Serabian*
> *Taunton, Massachusetts*

Chapter 4

April

Digging In:
Building Beds and Planting

When I go into the garden with a spade to dig a bed,
I feel such exhilaration and health . . .

—Ralph Waldo Emerson

Apple blossoms . . . asparagus spears . . . the first robin. Whatever you associate with April, you probably think *spring*. In nearly every region, gardens and gardeners shift into high gear this month—building beds, buying plants, and planting the garden.

With so much to do and so little spare time, you'll need tips and techniques that will help you garden smarter. Well, good news: They're coming right up!

Even if your garden is several years old, chances are you'll be building at least one new bed this season. For an easy-on-your-back way to make a bed, check out the technique on page 72. If you're like most gardeners, you'll also visit your local garden center several times this month to pick up bedding plants, row covers, tomato stakes, and maybe a shrub or two. While there, you'll undoubtedly peruse the bargain shelves. Before you put any nursery orphans into your shopping cart, though, be sure to read the advice on page 74.

Most of all, you'll be planting this month—not only your garden center goodies, but also all of those healthy organic seedlings you started a month or two ago. Look for plenty of planting tips starting on page 74, as well as for homegrown advice in the "Gardener to Gardener" section at the end of this chapter.

Gardener's To-Do List—April

**If you don't know what USDA hardiness zone you live in,
check the map on page 230 to find out.**

Zone 3

☐ When weather permits, set out new asparagus and rhubarb plants.

☐ As soon as the soil is warm and dry enough to cultivate, begin planting onions, peas, spinach, carrots, lettuce, beets, chard, and radishes.

☐ Late in the month, start tomatoes, peppers, and eggplants indoors.

☐ Direct-seed poppies, alyssum, bachelor's buttons, cosmos, and calendulas anytime (they don't mind chilly soil).

☐ Remove winter coverings from peonies, then top-dress them with compost.

☐ Set out new strawberries.

Zone 4

☐ Harden-off cabbage, broccoli, and cauliflower seedlings for a week before setting them out in the garden beneath cloches.

☐ Sow spinach, lettuce, and radishes, then cover them with plastic tunnels to get them growing in a hurry.

☐ Start tomatoes, peppers, and basil indoors.

☐ Late this month, plant potatoes and peas.

☐ Set out snapdragons, dianthus, and pansies.

☐ Divide daylilies, phlox, and other clumping perennials.

☐ Fertilize raspberries with compost and renew mulches beneath blueberries.

Zone 5

☐ Harden-off cabbage, then set out plants beneath cloches.

☐ Plant potatoes, peas, and spinach, followed by carrots, lettuce, and other greens.

☐ Start basil, tomato, and pepper seedlings.

☐ Plant new beds of asparagus and rhubarb.

☐ Set out new shrubs and trees.

☐ Fertilize established fruits with a thin layer of compost.

Zone 6

☐ Plant peas, potatoes, spinach, lettuce, and other leafy greens.

☐ Set out kohlrabi and broccoli under cloches.

☐ Between rainy spells, prepare beds for bush beans, sweet corn, and other summer vegetables.

☐ Weed flowerbeds, and remove protection from roses and other perennials.

☐ Dig up, divide, and transplant crowded daylilies, phlox, and hostas.

☐ Overseed sparse areas in lawns.

☐ Set out new fruit trees and shrubs, and fertilize established berries.

Zone 7

- ☐ Sow more carrots and lettuce early this month, and mulch potatoes with 6 inches of straw.
- ☐ Set out a few early-ripening tomato cultivars beneath cloches.
- ☐ At midmonth, sow sweet corn, cucumbers, summer squash, and bush beans, as well as herbs.
- ☐ Set out annual flowers, and plant roses, dahlias, lilies, and glads.
- ☐ Fill the backs of sunny flowerbeds with tall sunflowers or tithonia.
- ☐ Propagate groundcovers and hostas in shady areas.
- ☐ Set out blackberries and strawberries, and be sure to provide plenty of water.

Zone 8

- ☐ Check the brassica patch for cabbageworms; if you spot any, handpick them or apply *Bacillus thuringiensis* (BT).
- ☐ Set out tomatoes, peppers, and eggplant, and sow sweet corn, squash, cucumbers, and beans.
- ☐ Start sweet potato slips to transplant next month.
- ☐ Replace faded lettuce with okra, peanuts, field peas, or lima beans.
- ☐ Plant annual vines to shade heat-sensitive plants from summer sun.
- ☐ Top-dress roses with compost.
- ☐ Sow easy-growing annuals such as celosia, zinnias, and sunflowers.
- ☐ Late this month, thin peaches and plums to 6 inches apart.

Zone 9

- ☐ Replace spent spring crops with okra, asparagus beans, Malabar spinach, cherry tomatoes, and sunflowers.
- ☐ Cage tomatoes, peppers, and eggplant.
- ☐ Side-dress rows of garlic, shallots, and onions with compost.
- ☐ Late in the month, sow cantaloupe, pumpkins, and squash.
- ☐ Sow hot-weather flowers such as celosias, sunflowers, portulaca, and zinnias.
- ☐ Set out bedding plants, such as asters, coreopsis, impatiens, and salvia.
- ☐ Apply mulches!
- ☐ Harvest strawberries before slugs and snails beat you to it!

Zone 10

- ☐ Pull up bolted brassicas and leafy greens, and compost them.
- ☐ Try underplanting okra with sweet potatoes.
- ☐ Keep harvesting tomatoes, cucumbers, melons, and squash to encourage continued production.
- ☐ As parsley, coriander, and dill go to seed, replace them with plants of lemongrass, Cuban oregano, and Vietnamese coriander.
- ☐ Plant allamanda and other flowering vines as shade screens for decks, patios, and windows.
- ☐ Water and remulch beds frequently.

A Simple Way to Raise Your Beds

Whether you're planting food or flowers, your plants probably will perform best in raised beds. (Gardens in dry climates are the exception; see "Where *Not* to Raise Beds" on the opposite page.) Earlier planting and harvest, lush growth, higher yields, and easier maintenance are just a few of the many advantages of raised beds.

And if you've been procrastinating about raising your beds because you think the job is too labor intensive, rest assured that it doesn't have to require hours of backbreaking work. There are many ways to make a bed. Here's an easy method for gardeners of all levels. You don't have to remove any sod, you do very little digging, and you can plant right away.

1. Clear the area. If the sod in the area where you want to make the beds is full of persistent, deep-rooted weeds such as dandelions, remove them with a sharp knife or spade, being careful to remove as much of the taproot as possible. Otherwise, simply cut the grass or weeds to 1 inch high with a mower or trimmer.

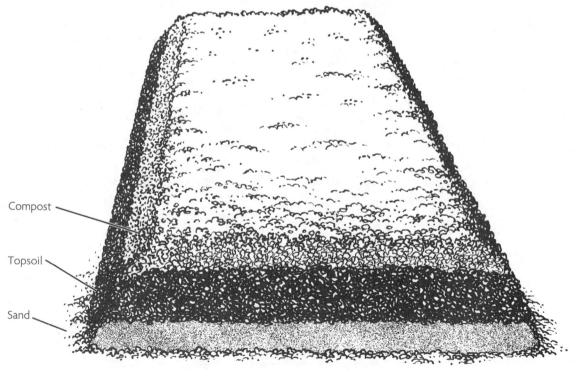

Compost

Topsoil

Sand

Existing soil line

2. *Define the beds.* Using stakes and twine, mark off the outline of each bed. For easy access, the width of the beds should be no more than twice the length of your arm (totaling about 4 to 5 feet); the length can be whatever you like. If you're making several beds in one area (as for a vegetable garden), remember to allow enough room for adequate pathways between the beds. Make main paths 6 to 8 feet wide and secondary paths 4 to 5 feet wide. That way, when the beds spill over with plants, the pathways never become too narrow to accommodate a wheelbarrow or people walking side by side.

3. *Build the base layer.* If the underlying soil is primarily clay or drains poorly, build a 6- to 8-inch-deep base layer for each bed, using sandy loam or builder's sand (coarse, gritty sand, available from home centers). Rake this layer to attain the desired depth and shape, adding graceful curves to ornamental beds. If your site is already blessed with good drainage, you can skip the sand base.

4. *Add topsoil.* Add a 6- to 8-inch layer of organic topsoil over the sand base, raking the soil for even coverage. If you aren't starting with sand, make the soil layer 8 to 12 inches deep.

5. *Mulch with compost.* Top-dress the beds with 1 or 2 inches of finished compost. This mulch layer suppresses weeds, conserves moisture, and slowly releases nutrients. Rake the surface so that it is level and smooth. Also rake the sides to create a gentle slope. Don't worry if the beds appear high; they will settle a few inches over the coming months.

6. *Cover pathways.* If the paths between your beds are already covered with sod, you can leave them that way and simply mow the grassy walkways to maintain them. Problem is, mowing the paths gets tricky when plants grow large and spill out over the beds. If you're comfortable with the location of your beds, you can create more-permanent pathways by using gravel or bark chips. First, cover the sod or soil with a heavy-duty barrier cloth (sold at most garden centers) to keep grass and weeds from growing through the path. On top of the cloth, spread at least 4 inches of the gravel or bark chips.

7. *Plant the beds.* When you're ready to plant, scoop out a planting hole and add a blend of compost and soil. Place each plant so that the top of the rootball is even with the surface of the new soil, then fill the hole with more of the soil blend. (Note that tomatoes usually are planted deeper to encourage large root systems.) Water each plant well.

Raised Beds: To Be or Not to Be

WHY RAISED BEDS?

Not sure you're ready to raise your beds this season? Consider these benefits:

♦ In wet climates or low-lying areas, raised beds allow water to drain quickly (so plants are less prone to root rot).

♦ The dry, well-aerated soil warms more quickly in spring, allowing earlier planting and harvest.

♦ Roots grow deeper, so plants have better access to nutrients and water.

♦ Foot traffic is confined to pathways, so soil in the beds doesn't become compacted.

♦ Plants can be spaced more tightly, increasing yield per square foot.

♦ After you build the beds, you can reuse them year after year with very little digging and cultivating.

WHERE *NOT* TO RAISE BEDS

If you garden in an area that is very dry or windy or where the soil is very sandy, you should avoid raised beds and choose *sunken* beds instead. Sunken beds retain moisture and are less prone to wind erosion. To make sunken beds, start by forming 3 × 4-foot mounds. Then rake the soil from the interior of each mound to the edges to form a basin with mounded edges. To retain even more soil moisture, cover the soil surface with mulch after plants are up and growing.

Buying Bargain Plants

Discount stores and supermarkets usually have plenty of bargain plants in late spring. Here's how to ensure that the bargain plants you're considering are truly good buys:

◆ Avoid plants that are yellow or wilted.

◆ Be sure plants are labeled.

◆ Examine the plants closely for insects (especially beneath leaves).

◆ Sniff the soil. If it smells bad, the plant may be dying.

◆ Never buy a sickly plant and expect to nurse it back to health.

◆ Don't buy a healthy-looking plant that's surrounded by sick plants. It probably has the same problem but isn't showing it yet.

Recipe for an Easy-Care Landscape

Imagine having beautiful flowerbeds and borders that need very little upkeep. The secret is to choose your plants wisely, then employ a couple of simple design tricks.

Shop Smarter

When shopping for plants, keep in mind these tips for a beautiful, low-maintenance landscape:

Choose plants that fit your site and soil type. For instance, if you have to deal with lots of shade, select plants adapted to that condition. Don't ask plants to perform unnatural acts, and don't try to control what you perceive to be their "bad" habits. Put each plant in its proper niche.

Use proven regional performers. Ask your local garden center or cooperative extension office for suggestions regarding which plants do best in your area.

Look for plants that have more than one strength. For example, instead of choosing plants that offer a couple of weeks of pretty flowers, go for plants that have beautiful blooms, interesting seedpods, and bright fall foliage. You should be able to enjoy your plants for more than just a few short weeks.

Splurge on a few larger shrubs and at least one good-size tree. These may cost $50 to $100 each, but nothing gives your garden a look of fullness faster than some mature, well-placed key plants.

Plant with Care

After you get your purchases home, provide the conditions that will help your plants thrive with minimal help from you.

Group plants in a naturalistic way. One way to give plants what they need is to group them in communities, the way they occur in nature. For example, use shade-loving understory shrubs and groundcovers beneath the canopy of a tree.

Give each woody plant room to develop. When planting shrubs and trees, consider their mature size and form, then give them the space they'll need to develop. If you prefer to space perennials more closely for immediate impact, be prepared to move them when they outgrow their positions.

Fill in gaps with easy-going annuals. Until perennials mature, you can make the area appear more lush by filling empty spaces with annuals. Give preference to native plants and those that self-sow their seeds. When the perennials mature, you can eliminate annuals entirely.

Planting Time

At last—the moment you've been preparing for! Now that your soil and beds are ready and the weather has warmed, it's time to begin planting. Whether you're sowing seeds or setting out transplants, both timing and technique are crucial to getting your plants off to a strong start.

Sowing Seeds Outdoors

Planting seeds directly into the soil is gardening the way nature does it. Before you begin sowing, check seed packages for recommended planting dates, temperatures, depth, and spacing. Timing is especially important. Some plants, such as peas, prefer cool weather; others, such as corn, will rot if planted before the soil warms up. (For suggested planting dates for specific crops, see the chart on page 80.)

You can choose either to plant seeds in straight rows (furrows) or to scatter (broadcast) them randomly over a wider area. Make furrows for large-seeded vegetables and cutting flowers by pulling the blade of a hoe or trowel along a length of twine strung between two stakes. For small-seeded vegetables, as well as flowerbeds and borders where you want a more natural-looking display, broadcast the seeds or scatter them in a wide band. Whether you use furrows or broadcasting, distribute the seeds sparingly so you won't need to thin later.

It's easier to plant seeds at the correct spacing than to thin them later. When sowing straight rows, such as for vegetable seeds, a notched board with a beveled edge makes a handy guide.

Taking Your Soil's Temperature

The most accurate way to time planting is to use a soil or compost thermometer (sold at garden centers and hardware stores). Take daily readings, starting at least 1 week before you expect to plant seeds outdoors. Brush away any surface litter or mulch and insert the thermometer to a depth of 3 inches. Every 5 days, add up your daily readings and divide by 5 to determine the average temperature. When the average soil temperature over a 5-day period reaches 45° to 60°F, plant cool-season crops. Sow warm-weather seeds when soil temperatures are between 65° and 80°F. Wait until the soil is consistently between 75° and 90°F before you plant hot-weather crops.

Planting by Nature's Signs

Let nature tell you when it's time to plant or watch for pests! Each year, both plants and pests develop in a predictable sequence that follows local weather patterns. Keep your own record of the development of both wild and cultivated plants, and note how these biological events coincide with changes in soil temperature and weather. For example, you might record in your garden journal the date when lilacs or bearded irises bloom, along with the accompanying temperatures. After a few seasons, the plants, insects, and wildlife around your home will help remind you when it's time to tend to your garden tasks.

Cover the seeds with fine soil (screen out any rocks with a sifter like the one on page 66) to a depth equal to three times the greatest diameter of the seed. Some seeds, such as lettuce, petunias, and begonias, require light to germinate. Lightly press these seeds onto the surface of moistened soil.

Always water gently after you sow seeds, taking care not to wash the seeds away. A fine, misty spray is best; you can buy a hose attachment at garden centers and hardware stores. Keep the soil evenly moist until you see stems and leaves popping above the ground.

Planting Transplants

Transplants give you a head start on the season and, if you're succession cropping, allow you to harvest more from a given amount of space. You can grow your own transplants from seed, as explained in Chapter 2, or buy them from a local nursery or garden center.

Hardening Off. Whether you've grown your own transplants from seed or are starting with store-bought plants, you'll need to "harden them off" before you plant them. Hardening off helps transplants make a smooth transition from indoors to outdoors. Start the process a week or two before you plan to transplant by watering less often, stopping plant feedings, and lowering the air temperature by several degrees. For plants growing in an open, undivided flat, cut into the medium between the seedlings with a knife to stimulate individual root growth and make it easier to untangle plant roots at transplant time.

Use a teaspoon or ice-cream stick to "prick out" seedlings. Handle them gently, holding onto a leaf and supporting the rootball rather than grasping the stems or delicate growing tips when moving plants from a seed tray and transplanting them into the garden.

After a week, move the plants outdoors each day for several hours during the warmest part of the day. Leave them in partial shade and away from strong winds, then bring them back indoors overnight. Gradually increase the time they're left outdoors. If you're away from home during the day, try short exposures early in the morning and again in the evening.

Planting Out. To get your plants off to the best start, water them thoroughly before you transplant. Some gardeners like to soak plants in compost tea or a weak solution of fish emulsion for about 15 to 20 minutes before planting.

To pop a transplant from its container, turn the pot upside down and tap on the bottom. Use your other hand to support the rootball as it slides out. Gently squeeze out plants growing in individual cells, and use a spoon or small stick to lift seedlings from seed trays.

Use a trowel or other hand tool to make a planting hole slightly wider than the rootball. Place the plant in the hole, gently spreading out any roots that are wrapped into a ball. Fill in with soil around each plant. Leave a saucerlike depression to hold water, but be sure the soil fully covers the block of potting mix around the roots. For tomatoes, peppers, and other plants that have a bare stem, set the rootball a bit deeper than it was in the pot to encourage more roots to develop.

If your plants are in newspaper or peat pots, tear away the collar that extends above the soil in each pot and use your finger to poke holes in the bottom of the pot, before setting the plant and pot into the garden. Be certain that none of the pot is sticking above the planting hole after you fill it in, or it will act as a wick and drain the moisture out of the rest of the pot.

When you're finished setting out your transplants, sprinkle the soil around each one with at least 1 quart of water.

Aftercare. Pamper your transplants for at least a week or two after you've planted them. Water whenever the soil dries out. Provide shade on bright days by covering the transplants with overturned baskets or shade cloth supported by hoops.

To protect plants from wind, use any kind of barrier that's handy: cloches made from half-gallon milk jugs with their bottoms cut away, or large open-ended juice cans sunk into the soil. A deep mulch of bulky organic materials, such as straw, can also help protect plants from wind. Plastic and fabric row covers (available at garden centers) are good for pulling over your plants on cold days and nights.

Remove the protective barriers when plants show signs of vigorous new growth or when variable spring weather gives way to more-stable summer conditions and warmth.

Bareroot Planting Basics

If you buy *bareroot* perennials, shrubs, or trees, you'll need to plant them as soon as possible. If you can't plant right away, remove the packaging and stick the roots in moist peat.

For slightly longer-term storage, heel-in plants. Dig a trench with one vertical and one slanted side in a spot sheltered from direct sun and wind. Lay the plants against the slanted side, and cover the roots with soil.

Before the plants break out of dormancy, uncover their roots and move them to their permanent site. When planting, dig a hole wide enough to spread out the roots, and deep enough so that the plant will grow at the same depth it was previously. Refill the hole with soil, water thoroughly, and mulch.

How Many Seeds Per Ounce?

Many mail-order catalogs tell you how many seeds are in a packet. But some seed packets—especially those that are sold in stores—give the weight of the seed enclosed but not their numbers. Use the list below to translate weight into numbers so you can estimate the amount you'll need to buy. If the weight of the contents is listed in grams, remember that 28.3 grams equals 1 ounce.

Vegetable	Seeds per Ounce	Flower	Seeds per Ounce
Beets	1,600	Blanket flower	15,000
Broccoli	9,000	Candytuft	9,500
Cabbage	9,000	Cleome	13,000
Carrots	23,000	Coneflower	40,000
Corn	120–180*	Cornflower	6,000
Cucumbers	1,100	Cosmos	4,000–5,000*
Eggplant	6,500	Globe amaranth	10,000
Lettuce	25,000	Larkspur	8,000
Lima beans	25–75*	Marigold	9,000
Onions	8,500	Mexican sunflower	3,000
Peas	90–175*	Morning glory	800
Peppers	4,500	Nasturtium	175
Pumpkins	100–300*	Periwinkle	20,000
Radishes	2,500	Poppy	260,000
Snap beans	100–125*	Portulaca	280,000
Spinach	2,800	Strawflower	45,000
Squash	120–400*	Sunflower	500–700*
Tomatoes	11,500	Zinnia	2,500–4,000*

*Some cultivars have larger seed than others.

Seedling Shelters

Harsh weather right after transplanting can damage even hardened-off seedlings. Protect transplanted seedlings from the elements with makeshift shelters, constructed with items found around the house.

To protect new transplants from strong sunshine, set up sun shields. Stick a shingle or board in the soil to block sunlight from one side, or completely cover seedlings with a newspaper tepee or a paper bag. Remove whole-plant covers after a day or two, or plants will become spindly.

Wind can knock over plants and draw water out of leaves faster than the limited roots can replace it. Wind blocks can help plants "weather" any windy conditions. Cut panels from milk jugs or milk cartons, then insert them near the plants on the side that gets the most wind (wind usually comes from the west).

If you're setting your seedlings out a little early or are expecting a late frost, you'll need to protect your plants from cold temperature. Cut the bottoms of plastic milk jugs or 2-liter soda bottles and anchor them to the ground around your seedlings by inserting a stick through the handle and anchoring the stick in the ground.

Throw an old sheer white curtain over garden plants to protect them from light frosts. They work just as well as many garden fabrics, and you can just toss them in the washer with your gardening clothes if they get too dirty.

Cut off the end of a clear plastic garbage bag, then slide the bottomless bag over your tomato cages to protect new transplants. If nights are still cold, you can gather the bag together at the top to retain heat, but be sure to leave the top open if the days are warm. Remove the bag once the weather has warmed.

Use under-the-bed, clear-plastic storage containers to keep the evening frost off new transplants. Be sure to uncover plants once morning comes.

Easy Insurance against Disease

You can easily protect your seedlings against *damping-off*, a fungal disease that rots seeds and seedlings, by blanketing them with a natural fungicide: milled sphagnum moss. Lightly sprinkle the moss over the surface of the seed-starting mix after you've planted your seeds. If the seeds need light to germinate (e.g., alyssum, ageratum, petunias, and snapdragons), sprinkle the moss on the soil mix first, then drop the seeds onto the moss.

Direct Seeding Dates

This table serves as a general timetable to help you schedule plantings out in the garden. But first, call your local extension agent to find out the frost-free date in your area. Use this date as your point of reference as you add or subtract weeks, depending on the hardiness of the crops you're planting. *Hardy* plants can withstand subfreezing temperatures. *Half-hardy* plants can withstand only some light freezing. The fruit and leaves of *tender* crops are injured by light frost, while *very tender* plants need warm temperatures (above 70°F) to grow. Any exposure to temperatures just above freezing will damage fruit and leaves.

FROST-FREE DATE _____

COOL-SEASON CROPS		WARM-SEASON CROPS	
HARDY: PLANT 4–6 WEEKS BEFORE FROST-FREE DATE	HALF-HARDY: PLANT 2–4 WEEKS BEFORE FROST-FREE DATE	TENDER: PLANT ON FROST-FREE DATE	VERY TENDER: PLANT 1 WEEK OR MORE AFTER FROST-FREE DATE
Asparagus	Beets	Beans, snap	Corn (depending on variety)
Broccoli	Carrots*	Corn (depending on variety)	
Brussels sprouts	Cauliflower		Cucumbers
Cabbage	Lettuce*	Okra	Eggplants
Collards	Potatoes	Tomatoes	Melons
Onions	Radishes*		Peppers
Peas	Swiss chard		Squash, summer
Spinach			Squash, winter

*These particular half-hardy plants can be planted outdoors at the same time as hardy crops if they are protected from extreme cold.

How to Figure the Last Planting Date

	Vegetable	Days to Maturity[1] +	Days to Germi- nation[2] +	Days to Trans- planting +	2 Weeks Short-Day Factor[3] +	before First Frost[4] =	Days to Count Back from First Frost Date
Frost- Tender	Beans, snap	50	7	Direct seed	14	14	85
	Corn	65	4	Direct seed	14	14	97
	Cucumbers	55	3	Direct seed	14	14	86
	Squash, summer	50	3	Direct seed	14	14	81
	Tomatoes	55	6	21	14	14	110
Survive Light Frost	Beets	55	5	Direct seed	14	—	74
	Cauliflower	50	5	21	14	—	90
	Lettuce, head	65	3	14	14	—	96
	Lettuce, leaf	45	3	14	14	—	76
	Peas	50	6	Direct seed	14	—	70
Survive Heavy Frost	Broccoli	55	5	21	14	—	95
	Brussels sprouts	80	5	21	14	—	120
	Cabbage	60	4	21	14	—	99
	Carrots	65	6	Direct seed	14	—	85
	Collards	55	4	21	14	—	94
	Radishes	25	3	Direct seed	14	—	42
	Spinach	45	5	Direct seed	14	—	64
	Swiss chard	50	5	Direct seed	14	—	69

[1] These figures are for the fastest-maturing varieties we could find. Fast-maturing cultivars are best for fall crops. But for the variety you have, get the correct number of days from your seed catalog.
[2] These figures for days to germination assume a soil temperature of 80°F.
[3] The short-day factor is necessary because the time to maturity in seed catalogs always assumes the long days and warm temperatures of early summer. Crops always take longer in late summer and fall.
[4] Frost-tender vegetables must mature at least 2 weeks before frost if they are to produce a substantial harvest.

Technique

Build a Simple Potting Bench

Planting doesn't happen only in the garden. Most gardeners also plant flowers, veggies, and herbs in pots for the deck or patio or to enjoy year-round as houseplants. This easy-to-build portable bench is perfect for all of your potting, transplanting, and seed-starting work. Because it's made from cedar, it looks good, weathers well, and is naturally rot-resistant. You can set up the bench wherever you need it—it's lightweight and easy to carry. When you're not using the bench, just fold it and hang it on a wall in a shed or the garage.

> ## MATERIALS
> 1" × 3" × 8' cedar boards (11)
>
> ⁵⁄₁₆" × 3" carriage bolts (4)
>
> 2" galvanized deck screws (40)
>
> ¼" galvanized deck screws (12)

1. Cut seven of the 8-foot 1 × 3 boards to the following lengths: eight 48-inch pieces; two 42½-inch pieces; two 20-inch pieces; three 18½-inch pieces; and two 10-inch pieces.

2. Form a 42½-inch × 20-inch rectangle by fastening two of the 18½-inch lengths between both 42½-inch pieces with the 2-inch screws; this is the apron.

3. Use the 2-inch screws to fasten the third 18½-inch piece between the sides in the center of the rectangle; this is the center batten.

4. Fasten the two 20-inch pieces across the ends of the rectangle, using the 2-inch screws at the ends and the 1¼-inch screws in the middle.

5. Fasten the 10-inch pieces in opposite corners of the rectangle, using the 1¼-inch screws; these are the spacers that allow the legs to fold properly. (See Figure 1.)

6. Fasten the eight 48-inch pieces to the top of the apron, using 2-inch screws, to complete the tabletop.

7. Turn the tabletop over and attach the 18½-inch center batten to it with 1¼-inch screws.

8. To make the legs, cut two 8-foot 1 × 3 boards into four pieces, 29½ inches long and tapering from 3 inches at one end to 1½ inches at the other. Round off the wide end and drill a ⁵⁄₁₆-inch diameter hole through the center point at that end.

Figure 1

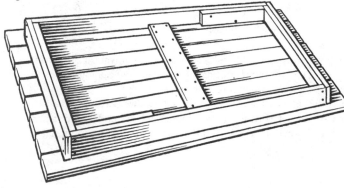

9. Drill four ⁵⁄₁₆-inch diameter holes through the apron and spacers, then bolt the legs to the tabletop. (See Figure 2.)

10. For leg braces, cut the remaining 8-foot boards into two 20-inch and two 17⅝-inch pieces. With the legs open slightly past 90 degrees, use 2-inch screws to attach the 20-inch braces to them so that the braces act as stops against the apron. (See Figure 2.)

11. Fold the legs up against the table and attach a 17⅝-inch brace across each pair of legs at the point where they just clear the opposite legs when folded; use 2-inch screws. (See Figure 3.)

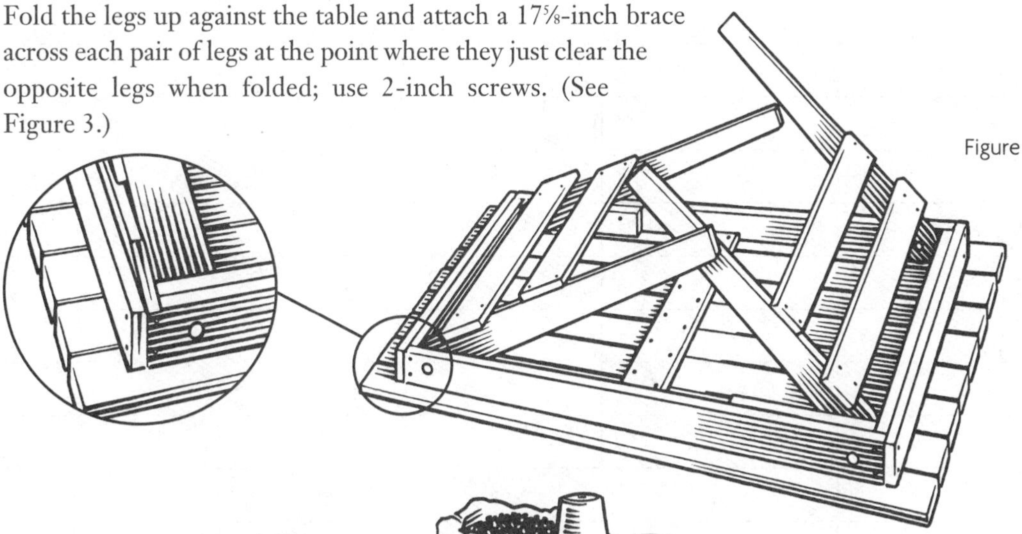

Figure 2

Figure 3

Vegetable, Flower, and Herb Log

Plant Name	Date Planted	Notes

GARDENER TO GARDENER

No-Dig Beds

Last spring I wanted to expand my garden to accommodate some new cucumbers and melons, but I didn't have the time to dig new beds. So I decided to try growing the crops in some old straw bales I had on hand. The experiment worked so well that I may create an entire garden from straw bales. Here's what I did:

1. Using a crowbar, I punched holes into each bale, then sprinkled several cups of bloodmeal fertilizer over each bale. For the next couple of weeks, I watered the bales with a tea of manure and alfalfa meal.

2. Next, I used a serrated knife to cut two holes about 5 inches around and 8 inches deep into each bale.

3. I then filled the holes with a mixture of compost and soil, and planted the cucumbers and melons in the holes. I was amazed at how fast the vines grew, and the bales held enough moisture that I seldom had to water them.

Cheryl Webb
Houlton, Maine

Board Prevents Soil "Crusting"

Carrot seeds need to be kept moist until they germinate, which can take several weeks. If you water by hand during that time, the soil often forms a crust that seedlings cannot penetrate.

Here's how to keep the soil moist with just a single watering before you sow the seeds. Dig a narrow, shallow furrow and water it well. Sprinkle your seeds into the furrow, then cover them lightly with soil. Lay a 4- to 6-inch-wide board (untreated wood) over the row. Check under the board every couple of days; as soon as the first sprouts appear, remove the board. The soil will stay moist without getting crusty and won't dry out no matter how hot it gets.

Raymond Brilhart
Tidioute, Pennsylvania

Hoe Sows Seeds

Here's an easy and quick way to plant seeds at the correct depth in the garden. Just paint three different-colored stripes on the end of your hoe handle (I used blue, red, and yellow), placing them ¼ inch, ½ inch, and 1 inch from the end. When it's time to plant, simply push the handle into the soil until you reach the required depth for that type of seed, then drop the seed into the ground. Use the blade end of the hoe to gently cover the seeds after they're planted.

Craig Grodman
New Hampton, New York

All in the Timing

Perhaps you're familiar with the old-time garden advice to "plant your eggplants when the roses are in bloom" because the soil will be sufficiently warm by then. Well, my grandmother, who was very observant, had a timing tip you might not have heard of before: "Plant your hollyhocks when the white oak leaves are as big as squirrels' ears."

Helen White
Ocean Shores, Washington

GARDENER TO GARDENER

Storm-Proof Trellis

Year after year, summer storms would come along and knock over the trellis on which I grew morning glories, uprooting all of the plants at the same time. But last year I came up with a new, easy-to-make trellis that solved the problem.

First I screwed some cup hangers (available at most hardware stores) into the side of my shed at the very top and painted them to match the color of the shed. Then I cut several metal coat hangers into U-shaped pieces, and stuck them upside down into the ground about a foot away from the shed. I ran twine from the cup holders to these wire anchors (beginning with the holder at the upper left corner), then back up again, and so on, until the trellis was complete. Last year we had several bad storms, but my trellis and morning glories weathered them all beautifully!

Joan Rousseau
Dudley, Massachusetts

Sow Seeds Standing Up

If your back hurts from bending over to plant seeds, try my method instead. Simply plant the seeds through an aluminum pipe or any lightweight tube. First make a furrow with a hoe, then set the pipe in the groove and drop the seeds through it. I use an apron to carry my seeds and plant row after row this way.

Robert Southam
Nazareth, Pennsylvania

Superior Cloches

Vegetable transplants usually get off to a better start in the garden if you use cloches to protect them from cold winds and insect pests. But I find that traditional hotcaps, such as milk jugs, are too short to protect seedlings for more than a week or two.

Instead, I use the cone-shaped wire tomato cages sold at garden centers. I turn them upside down and wrap the sides with clear plastic. I then cut off the long wire legs and bend the cut-off pieces to create U-shaped stakes to secure the inverted cages to the ground. Finally, I cover the tops of the tomato cages with pieces of row cover—this keeps out bugs and prevents overheating in the mini-greenhouses.

Charles Nardo
Egg Harbor Township,
New Jersey

Chapter 5

May

Cultivating Vegetables Organically: Peppers

I love spring anywhere, but if I could choose,
I would always greet it in a garden.

—*Ruth Stout*

For at least a week or two, those transplants and seeds you planted last month might appear to do nothing at all. You'll check your garden daily (or more often) for signs of growth—some indication that you've done things "right." But the soil will remain barren and lifeless.

Then, just when you've decided you must have planted too deeply, or that some thief in the night made off with your potential treasures, those beds come alive. All at once, it seems, the world is green and growing by leaps and bounds. Blossoms open. Birds sing. And the fragrance of spring-blooming bulbs fills the air. And you, too, feel more alive—because you've played a part in making it happen.

Chances are, at least some of those suddenly growing garden plants are peppers. With their amazing diversity of colors, shapes, and flavors, peppers are particularly rewarding to grow. And, as you'll soon read, you can bring in bushels of delicious, healthful peppers no matter where you garden. The secret is to choose the right peppers for your region, then employ a few simple cultivation techniques.

Gardener's To-Do List—May

**If you don't know what USDA hardiness zone you live in,
check the map on page 230 to find out..**

Zone 3

- [ ] Plant peas and potatoes.
- [ ] Harden-off cabbage-family seedlings and set them out at midmonth.
- [ ] Direct-seed lettuce, spinach, and root crops.
- [ ] Plant asparagus, rhubarb, and celery.
- [ ] Fill flowerbeds with pansies, cosmos, and petunias.
- [ ] Set out new berry plants.

Zone 4

- [ ] Plant peas and potatoes, and direct-seed parsley and kohlrabi.
- [ ] Get busy sowing lettuce, carrots, beets, and radishes.
- [ ] Start pumpkins indoors for planting next month.
- [ ] At month's end, sow beans, sweet corn, muskmelons, cucumbers, and squash; set out tomato plants under cloches.
- [ ] Plant new perennial flowers and sow seeds of cool-weather annuals such as nasturtiums and cosmos.
- [ ] Mulch fruit trees with compost.

Zone 5

- [ ] Plant peas and potatoes right away, followed by beets, lettuce, radishes, and spinach.
- [ ] Sow carrots and early cukes around midmonth.
- [ ] Toward month's end, plant corn, beans, muskmelons, squash, watermelon, basil, tomatoes, and peppers.
- [ ] Fertilize broccoli now to keep it growing fast.
- [ ] Thin leafy greens.
- [ ] Plant sunflowers and other colorful summer annuals.
- [ ] Plant roses and perennials by midmonth.

Zone 6

- [ ] Harden-off tomato and pepper seedlings for transplanting in midmonth.
- [ ] Sow basil, sweet corn, beans, cucumbers, squash, and melons.
- [ ] Plant leaf lettuce in the partial shade of taller plants.
- [ ] Thin leafy greens and use them in salads.
- [ ] Mulch potatoes.
- [ ] Top-dress roses with compost.

- [ ] Sow sunflowers, zinnias, marigolds, and other summer annuals.
- [ ] Wait for the fading foliage of spring-blooming bulbs to turn brown before you remove it.

Zone 7

- [ ] Set out the last of your tomatoes, peppers, eggplants, and sweet potato slips.
- [ ] Be sure to harvest leafy greens often because they soon will bolt.
- [ ] Plant sweet corn, beans, melons, gourds, okra, and field peas.
- [ ] Keep mulching potatoes.
- [ ] Have *Bacillus thuringiensis* (BT) ready to apply at the first sign of caterpillar pests.
- [ ] Plant summer-blooming bulbs, such as cannas, glads, and tuberous begonias.

Zone 8

- [ ] If you expect a dry summer, install drip irrigation lines.
- [ ] Mulch everything!
- [ ] Stop watering onions, garlic, and shallots when the foliage begins to turn yellow.
- [ ] Set out sweet potato slips on a cloudy day.
- [ ] Plant okra, eggplant, peanuts, cherry tomatoes, field peas, and limas.
- [ ] Reseed sparse patches of lawn.

- [ ] Dig up and divide irises, daylilies, and Oriental poppies after they finish blooming.
- [ ] Thin fruits after the first "extras" fall from trees. Water all fruit trees and bushes during dry spells.

Zone 9

- [ ] Harvest remaining cool-season crops.
- [ ] Thin early sowings of corn and beans.
- [ ] Plant okra, melons, black-eyed peas, and jicama.
- [ ] Mulch beds if you haven't already done so.
- [ ] Begin pinching the tips of chrysanthemums and impatiens to encourage bushy growth.
- [ ] Fertilize late-blooming crape myrtles.
- [ ] Keep flowers, lawns, and roses well watered.
- [ ] When fruits begin to bloom, give them a booster feeding of fish emulsion.

Zone 10

- [ ] Plant only heat-loving edibles and ornamentals now, and water them often.
- [ ] Protect developing tomatoes and peppers from blistering and cracking by providing some shade.
- [ ] Keep tomatoes and peppers evenly moist.
- [ ] Remove faded flowers from annuals and perennials.
- [ ] Be sure fruit and nut trees receive enough water to prevent excessive fruit drop.

6 Steps to Productive Peppers

Planning to grow peppers this season? Great! Peppers are chock-full of good flavor and nutrition. Here's a step-by-step guide to help you reap your best pepper crop ever, whether you're starting with your own transplants or planting ones you bought at your local garden center.

1. Choose and prepare the site. The right site can make all the difference in how well peppers perform. Choose a sunny, well-drained spot where peppers haven't grown recently. The soil should be deep, rich, and loamy. If yours isn't, amend it with about 1 inch of compost. Avoid adding too much nitrogen to the soil, however. Excessive nitrogen can cause the plants to grow too fast, making them more susceptible to disease and *less* productive.

2. Harden-off the seedlings. Before you plant your pepper seedlings, you'll need to harden them off by gradually exposing them to outdoor conditions. (This gradual exposure to the weather helps seedlings adjust, so they'll be less stressed when you plant them. And less stress means bigger, more productive plants.) When daytime temps reach the mid-60s, set the seedlings in a sheltered location outdoors, such as next to the house or garage, for a few hours each day for 3 or 4 days. Over the next week, slowly extend that outdoor time. Meanwhile, as the seedlings are becoming accustomed to the outdoors, you can warm the pepper bed by covering it with black plastic mulch or dark landscape fabric.

3. Plant the peppers. Peppers like warmth, so wait to plant until nighttime temperatures have consistently reached 60°F and all danger of frost has passed. If possible, set out your peppers on a cloudy day to help reduce stress on the plants. Space the plants 12 to 20 inches apart, depending on the mature size of the variety, and set them a bit deeper than they were in their containers. (Like tomatoes, peppers grow extra roots from the buried portion of the stem.) Stake or cage taller varieties so that the stems do not break in strong winds or due to a heavy fruit load. After you plant the seedlings, water them well.

4. Water and mulch. Throughout the growing season, make sure your pepper plants receive at least an inch of water a week. Check the peppers often during periods of extreme heat and drought, when each plant can easily take a gallon of water a day. If you live in a very hot, arid region, add a thick layer of organic mulch to help retain soil moisture and to help moderate the soil temperature. But do this only *after* your soil has warmed—mulching cool soil will keep it too cool and stunt the plants' growth.

5. Pinch off the first flowers. As difficult as it might be for you, pinch off any early blossoms that appear on your pepper plants. This won't harm the plants. In fact, it helps them direct their energy into growing, so you get lots of large fruits later in the season (and a higher overall yield) instead of just a few small fruits early on.

6. Reap the bounty. You can harvest the peppers at their immature green or purple stage, but the flavor will be sweeter if you wait for them to turn their mature color—usually red, but sometimes golden yellow or orange. Italian fryers, jalapenos, and Cubanelles are possible exceptions: Many people prefer the flavor of these peppers when they are full size but still green. To harvest the peppers, cut them off with hand pruners. Pulling them off by hand can damage the plant.

Pretty Peppers for Small Spaces

No room to grow big, sprawling pepper varieties? Grow *ornamental peppers* in pots on a deck or patio, or tuck the plants into a sunny flowerbed for a bright, edible landscape. For tasty, brightly colored peppers on compact plants, try these easy-to-grow varieties:

SWEET
'Cherrytime'
'Jingle Bells'
'Sweet Pickle'

HOT
'Filus Blue'
'NuMex Twilight'
'Peruvian Purple'
'Poinsettia'
'Super Chile'
'Thai Hot Ornamental'

A bountiful harvest of peppers makes a great table centerpiece.

Sweet Peppers to the Max

It's impossible to have too many ripe sweet peppers. You can slice, stuff, stir-fry, and sauce these tasty garden treats all summer long, and you'd still want a big batch left over to freeze for use throughout the winter! The secret to growing those ultrapecks of sweet peppers is to plant high-yielding varieties and practice bounty-boosting techniques.

Peerlessly Productive Peppers

If you're fortunate enough to garden in an area with a long growing season and moderate climate, you can grow big, blocky *bells* like 'Jupiter' and 'Park's Whopper Improved'. In university research trials, both of these varieties distinguished themselves as superproductive—'Jupiter' scored big in California tests, while 'Park's Whopper Improved' was one of the very best performers in Florida trials.

But if your conditions are less than ideal, you'll do better with *non-bell* sweet peppers like these:

◆ Small-fruited peppers (3 inches or less across), such as 'Ace', 'Jingle Bells', and 'Northstar'. Unlike large-fruited varieties, these hold onto their flowers and continue to set and produce plenty of small- to medium-size peppers regardless of the temperature.

◆ Wedge-shaped sweet peppers, such as 'Gypsy'. Even in areas with long, cold springs, 'Gypsy' produces at least a dozen ripe, red peppers per plant. And in milder areas, that yield can easily triple.

◆ Long, pointy "frying" peppers, such as 'Cubanelle' and 'Sweet Banana'. While big bells take a midsummer siesta, frying peppers continue to pump out lots of sweet, thin-walled fruits.

Have a short season and/or poor conditions and still feel you *must* grow bells? Try 'Vidi'. No matter what the season's stresses, each plant bears plenty of 6- to 7-inch elongated bells, weighing up to half a pound each.

Get Out and Get Growing

Choosing productive varieties is only part of the game. You need to get your plants up and growing quickly because most peppers require a long, warm season to come to fruition. (For information on how and when to start pepper seeds, see Chapter 2.)

Remember, though, that peppers thrive in warm conditions. Don't jump the gun by setting out your transplants too early—wait until all danger of frost has passed and the soil has warmed. In cool climates,

Gold and Chocolate?

Are you one of the many gardeners who treasure the super-sweet flavor and golden glow of peppers that turn yellow—rather than red—when ripe? Then you must grow 'Orobelle'. It produces loads of large, juicy peppers that ripen up fast, even in climates not conducive to growing most peppers.

Chocolate-colored peppers generally don't yield as much as red or yellow types, but gardeners love to grow them for their unique color and super sweetness (though not chocolate flavor; sorry). High-yielding "chocolates" include 'Sweet Chocolate', with small, narrow fruits that quickly ripen to rich, deep brown, and 'Chocolate Beauty', which takes a couple weeks longer to ripen but develops a flavor that's worth the wait.

where the soil takes a long time to warm up in spring, you can use black plastic or black fabric mulch to help thaw out the ground quickly. (See "Landscape Fabric Warms Soil and Blocks Weeds" on page 95.)

After you've hardened-off your seedlings, plant them in a sunny, fertile site. If your soil is sandy and porous and you get a lot of rain, amend the planting holes with plenty of compost, then side-dress the plants in midseason with more compost or a balanced organic fertilizer. If you have heavy clay soil, simply work a 2-inch blanket of compost into the pepper bed at planting time. Don't add nitrogen to the soil unless a soil test says it's necessary—excess nitrogen can lower the plants' yield.

When planting your seedlings, be sure to go deep! Just like tomatoes, pepper plants yield more when planted deeper. Studies in Georgia and Florida found that peppers planted with their first true leaves just above the soil line produced 10 to 15 percent more fruits than plants placed at the standard depth.

Above ground, give your plants plenty of space. According to another Florida study, peppers planted 20 inches apart produced nearly three times as many fruits as plants spaced the standard 10 inches apart. That adds up to about 50 percent more peppers from the same amount of space!

More Tips for More Peppers

Which mulch is best for producing prolific peppers? Wheat straw is a good bet. In a 3-year study at Oklahoma State University, pepper plants mulched with a 6-inch layer of wheat straw consistently yielded fruits that were more plentiful, bigger, and less blemished than those mulched with black or white plastic or a "living mulch" of rye. Avoid mulching your peppers with leaves. Researchers in Connecticut found that leaf mulch actually *reduced* pepper yields because the leaves lowered the soil temperature by as much as 10°F.

After your peppers are warm and cozy in the ground, watch for the first flowers that appear on your plants—then pluck them off! A Massachusetts study found that pepper plants yield more over the season if the first flowers are removed. Removing the flowers causes the plants to develop additional branches on which more fruit can grow. The way in which you water your plants will influence yield, too. In a Louisiana study conducted during an unusually dry year, plants watered by sprinklers produced twice as many peppers as those watered by drip irrigation systems. The researchers concluded that the water from the drip lines percolated through the same areas of soil, leaching nutrients away from the plants' roots, while the sprinklers distributed moisture

Plant pepper seedlings slightly deeper than they grew in their containers to encourage a strong, healthy root system and higher yields.

more evenly. Here's a bonus: Sprinklers also help when flowering peppers are stressed by hot weather. According to a Cornell University plant scientist, a solid sprinkling for 30 minutes at midday and again in midafternoon "can lower temperatures to safer levels."

If you live in an area where heat stress lasts for weeks rather than hours, a shade cloth may help you pick more perfect peppers. In the same Oklahoma State study, white row covers were spread over hoops above bell pepper plants and left open on the sides for ventilation; the plants produced more and bigger fruits, and were protected from sunscald injury. Shade cloth offers the added advantage of protecting plants from aphids that spread deadly pepper viruses. Growers in Israel—where viruses are epidemic—found that white cloth deterred aphids from landing nearly 100 percent of the time.

In places where "summer" lasts longer than 6 months, you can boost your pepper yields by doing what commercial growers do—plant peppers twice, once in late winter and again toward summer's end. Or, prune back the plants in midsummer after they've ripened their first batch of peppers, and give them plenty of water and fertilizer. You'll be rewarded with a second, exceptionally large crop at the end of the growing season.

Don't Crowd 'Em!

When starting peppers from seed, be sure to give seedling roots plenty of room to roam! If you don't, crowded conditions could mean that you'll end up with far fewer peppers.

A researcher at a Griffin, Georgia, experiment station found that pepper seedlings with crowded roots produced fruits on existing branches instead of growing more branches, which would have produced an increased yield. By the end of the season, the plants that were crowded early in life produced far fewer peppers than those that weren't. To avoid this yield-compromising situation, transplant your pepper seedlings into roomy 3-inch pots about a month before you plan to plant them to the garden.

Staggered Harvest Spurs Production

Get more from your bell pepper plants by picking some of the fruits in the green, immature stage, then leaving the rest on the plant to color up and mature. Here's why: Peppers commonly form their first fruits in a center cluster. If you allow all of these fruits to ripen, they are susceptible to rotting due to the moisture trapped between them. You discourage the plant from setting more fruit if you leave the first flush of peppers untouched, so remove at least half the fruits. For green-pepper lovers, picking the first fruits is a bonus because they're larger than later ones.

Grow Red Hot Chile Peppers!

Want to have a real culinary edge on your neighbors? Grow a flavor sensation that you can get only from your own garden—hot peppers. Despite what you might have heard, you can grow these fiery little fruits almost anywhere if you start with varieties suited to your region.

Northern Picks

To ensure a full harvest of peppers before frost, northern gardeners must select varieties that mature in 90 days or less (50 to 75 days is ideal). For productivity and dependability, experts recommend the following types and varieties:

♦ **Northeast:** Hungarian wax peppers (especially early-maturing, open-pollinated varieties); ancho peppers (also known as poblanos); and early-maturing jalapenos, such as 'Early Jalapeno'.

♦ **Upper Midwest:** Cayenne peppers (especially 'Ring of Fire' and 'Golden Cayenne'); 'Szentsi Cherry' (super hot); and 'NuMex Joe E. Parker' (mild).

♦ **Northwest:** 'Arledge Hot' (hot, conical peppers); 'Ring of Fire' cayenne; Hungarian wax peppers; and, where summers are hot, Anaheim peppers.

Southern Selections

Southern gardeners don't have to be as fussy about which pepper varieties to choose because peppers love that long, hot growing season. But be sure to include at least a few of the following highly recommended types and varieties:

♦ **Southeast and Gulf Coast:** Cayenne peppers (especially 'Carolina Cayenne' and 'Super Cayenne'); jalapenos ('Jalapeno M', 'Jalapa', and 'Early Jalapeno'); tabasco peppers; serrano peppers; 'Bolivian Rainbow' (multicolored edible/ornamental fruits); and 'Chile de Arbol' (medium-hot fruits great for drying).

♦ **Southwest and California:** All of those listed for Southeast and Gulf Coast, as well as bird peppers (especially chile pequin, which is disease- and pest-resistant and tolerates extreme heat).

Landscape Fabric Warms Soil and Blocks Weeds

Many gardeners use black plastic mulch to warm their soil in spring before setting out pepper transplants. But black landscape fabric (sometimes sold as "weed barrier") is a better option. Unlike black plastic, the fabric allows air and water to reach plant roots. And soil doesn't overheat the way it can with plastic.

A week before planting peppers, market gardener Janette Ryan-Bush of Iowa City, Iowa, lays down strips of landscape fabric to warm the soil, leaving a 3-inch gap between the strips for planting. In the warm soil, the plants quickly grow large enough to shade out any weeds that might grow in the gaps.

Hot Pepper Vinegar

What to do with all those hot peppers? Turn your surplus into hot pepper vinegar for your pantry or for holiday gift giving. Here's how:

First, put a few of your prettiest whole (clean) peppers into clean jars. Put equal amounts of chopped peppers and vinegar into a blender and whiz them up. Pour the resulting liquid into a pot, bring it to a boil, then simmer it for 30 minutes. Strain the liquid, then pour it over the whole peppers in the jars.

The King of Sting

And what about the legendary habanero—the chile pepper said to be the hottest of them all? Habaneros (and the related Scotch bonnet peppers and datils) require a relatively long growing season with warm, humid conditions, such as those found in the Southeast. If you can provide those conditions, you'll find habaneros well worth growing. (They do not perform well in arid regions.)

For the true culinary thrill seeker, habaneros are the king of sting. Although they are blisteringly hot, the burn is short-lived. After you recover from the sting, you'll find that habaneros have an unmistakable flavor that is likened to citrus and sometimes to apricot.

Immature habanero chile peppers are green, while ripe ones turn a brilliant orange, yellow, red, brown, or even white. And you don't need many plants to fill your hotter-than-Hades needs—with habaneros, it's very true that "a little goes a long way."

Get Ready, Get Set . . .

Hot peppers are relatively easy to grow once you get the seeds started. Hot pepper seeds can be very slow to germinate—especially the varieties that haven't been in cultivation for very long. To break the dormancy of these wilder types, soak the seeds for 3 or 4 hours in room-temperature water before you sow them.

The seeds will germinate fastest in soil that's 85°F, which you can achieve by using a waterproof heating mat. If your soil is cooler than that, you can speed germination by soaking the seeds in salt water (1 tablespoon of salt per quart of water) for 2 or 3 days, says a Texas A&M University researcher. Be sure to rescue your seeds from this salt solution when the time is up, or you could stymie germination altogether.

After the soak, put the seeds on moist paper towels, stick the towels in a plastic bag, and put the bag in a warm place (such as on top of the refrigerator or on a counter above a dishwasher). Check inside the towels frequently. As soon as you see sprouts, plant them in peat pots or small containers and set them beneath bright lights.

Feed the seedlings lightly every couple of weeks, and in about 6 to 8 weeks, they should be about the right size to transplant to the garden. Harden-off the young plants before you transplant them, about 2 weeks after your last frost or when soil and air are consistently warm.

Keeping Them Going

So your peppers grow and prosper, and you pick and pickle and dry and salsa until the end of the season. Should you cover your plants and try and eke a few extra fruits out of them? No, says the Chile Institute in

Las Cruces, New Mexico. If you bring a plant indoors and replant it next year, it can yield another season of peppers.

Small-fruited types, such as bird peppers and habaneros, are the best candidates for overwintering indoors. You may even get some ripe fruits to enjoy through winter! This technique also allows northern gardeners to harvest long-season peppers that ordinarily won't ripen in a single, short growing season.

To keep a pepper plant growing and producing year after year, dig it up at the end of the season, then prune it lightly. Prune away old anchor roots, but leave intact the shallow surface roots, which supply food to the plant. Also prune back the top of the plant so that it is about the size of the pruned roots. Don't worry about how the plant looks, though. Your main goal is to keep the plant alive until next spring.

Pot the plant, then reverse the hardening-off process by bringing the plant indoors for successively longer periods over a week or two. After the reverse hardening off, put the plant in a sunny window or beneath fluorescent lights. Keep it away from heating vents or other places where it could easily dry out, and don't set it right up against a cold window.

In spring, when you normally set out pepper seedlings, move the potted pepper outdoors, taking care to harden it off just as you would seedlings. You can leave it in the pot, or stick it back in the ground. Either way, you'll have hot peppers before the first bell on the block turns ripe!

Salsa Kit Delivers Fun and Flavor

Brighten a friend's day with this twist on gifts from the garden: Give him or her a make-your-own-salsa kit, complete with homegrown ingredients and this salsa recipe:

1 cup seeded, diced tomatoes
2 tablespoons diced onion
1 or 2 garlic cloves, minced
½ teaspoon sugar
Salt, to taste
1 seeded, minced jalapeño or serrano chile pepper
1 seeded, minced sweet yellow pepper
Chopped cilantro, to taste

Keep peppers growing and producing year after year. When weather turns cool, dig up a plant, prune it lightly, then pot it and bring it indoors. Return the plant to the garden again in spring, after the weather warms.

Combine the ingredients and mix well. Let the salsa stand for 30 minutes before serving.

Top 10 Most Common Pepper Pests and Diseases

When Peter Piper planted his peppers, he may have plotted a plan against pests and pestilence in one of these ways.

1. **Blister beetles:** This striped beetle is both friend and foe; adults feed on pepper plants and other vegetables, but their larvae eat grasshopper eggs. Knock adults from plants into soapy water, but wear gloves because contact with crushed beetles blisters skin.

2. **Cutworms:** The nocturnal cutworm caterpillar attacks plants by curling around seedling stems and eating through them. Use cutworm collars and apply beneficial nematodes to the soil.

3. **Tomato hornworms:** The bright green larvae of this moth species chew large holes in leaves and may completely strip young plants. Handpick caterpillars and then drop them into soapy water.

4. **Pepper weevils:** These ¼-inch-long, dark gray insects lay eggs in pepper buds or fruits. The larvae leave dark cavities in the spongy inner tissues of pepper fruits. Invite natural predators such as birds and wasps to dine on these pests.

5. **Root-knot nematodes:** Microscopic soil-dwelling worms cause plants to wilt and lack vigor. To eradicate nematodes, grow a cover crop of marigolds or rye in infested areas and turn.

6. **Blossom-end rot:** A lack of calcium leads to the development of this disorder, which creates dark brown or black spots on the bottoms of immature fruits. Keep plants evenly watered to ensure a steady flow of calcium to the plant, especially while fruit is forming.

7. **Damping-off:** Seedlings that suddenly fall over and rot are most likely affected by damping-off. Prevent this problem by keeping the soil in which seedlings grow slightly dry and by not over watering.

8. **Bacterial spot:** This bacterial disease causes purplish gray spots on upper sides of leaves and raised ones on the backs of leaves. Pull up and dispose of infected plants.

9. **Viruses:** Mosaic is the most serious viral disease for peppers. Leaves become narrow and thickened, and they appear stringy. Pull and dispose of severely affected plants.

10. **Southern blight:** This fungal disease is most common in warmer climates. Mulch with compost after planting peppers, and rotate plants to avoid planting peppers in the same plot 2 years in a row.

Preserving Your Peppers

After you harvest those peppers you've worked so hard to grow, use them in these terrific recipes that bring out the flavor of your produce.

Pepper-Onion Relish

4 cups finely chopped onions

4 cups vinegar

2 cups finely chopped green peppers

2 cups finely chopped sweet red peppers

½ cup honey

Combine all the ingredients in a large enamel or stainless steel pot, and bring to a boil. Cook for 45 minutes, or until slightly thickened, stirring occasionally.

Pack the hot relish into hot, scalded pint jars, filling to the top of the jar. Seal tightly. Cool and store in the refrigerator.

Makes 4 pints.

Pickled Green Peppers

3 pounds green peppers, cleaned and sliced lengthwise

1 quart apple cider vinegar

¼ cup honey

Steam the pepper strips for 2 minutes. Drain. Combine the vinegar and honey in an enamel or stainless steel saucepan. Bring to a boil.

To can: Pack the pepper strips in hot, scalded pint jars. Cover with the hot vinegar mixture, leaving ¼-inch headspace. Seal and process for 10 minutes in a boiling-water bath.

Makes 4 or 5 pints.

Solutions

Pepper Problem Solver

Q: For the past few years, a fat worm about ½ inch long has been eating the pithy parts inside my peppers, leaving a brown slimy mess. I can't see any entry holes. What is this pest and what can I do about it?

A: Sounds like pepper maggots. They get inside peppers when a yellow fly with clear, banded wings lays its eggs just below the skin of the young fruit. When the eggs hatch, the larvae move inside to feed. Often, you don't notice any damage until the peppers turn prematurely red and begin to rot.

To disrupt the pest's life cycle, remove and destroy infected peppers before the maggots leave them to pupate in the nearby soil. For added insurance, cultivate the soil in fall to kill any pupae that may be present. Also weed out alternate hosts, such as horse nettles and ground cherries, in or near your garden. Next season, use lightweight row covers to shield young plants from the flies.

Q: Help! My bell peppers look pretty but they taste bitter. I feed them with composted manure. What am I doing wrong?

A: A couple of things could be going wrong. Excess nitrogen (from the manure) can cause bitterness. If your plants are exceptionally tall and vigorous and have very deep green leaves, they're probably getting more nitrogen than they need. Next year, skip the manure and other nitrogen-rich fertilizers, such as bloodmeal and fish emulsion.

Another possibility is that your peppers are exposed to too much intense sunlight. If you garden in a hot, cloudless region, use shade cloth (sold at garden centers) to give your peppers some heat relief. When the fruits begin to develop, erect a simple frame over the plants and drape the cloth over it. (Don't shade the peppers too soon, though, or you'll risk stunting the plants' growth.)

Q: How can I get my green peppers to turn red? I haven't had any luck even with fast-maturing cultivars like 'Northstar'.

A: Be sure to start seedlings early enough and get them into the garden as soon as the soil and air are warm enough. Keep in mind that the number of days to maturity listed in the catalogs usually refers to days from transplant to the first mature *green* pepper. You have to wait a few more weeks beyond that to get the first ripe, red peppers. Even early-maturing varieties need about 3 weeks to ripen from green to red.

You might also want to designate a few plants as "red-pepper plants." Don't harvest any green peppers from these plants. By not picking them, you force the plants to push along the ripening process.

Q: Last year I got a bumper crop of hot peppers that I stored in the freezer after harvesting them. But when I used them, I found they weren't as hot as the previous year when I grew the peppers in a poor site and gave them little water. Did the freezing make them milder or was it the growing conditions?

A: It was the growing conditions: the worse you treat hot pepper plants, the hotter the fruits will be. Stress—too little food and water—causes hot pepper plants to turn up the heat. Freezing or cooking doesn't affect the heat of hot peppers, say experts. In the kitchen, the only way to lessen the heat of a hot pepper is to trim out the white, pithy area inside—that's where the heat is most concentrated.

Q: My pepper plants don't produce more than a couple of peppers. It seems that the blossoms fall off before any fruits can form. How can I keep this from happening?

A: Don't rush to plant your peppers outdoors too early. Peppers require warm growing conditions and are very sensitive to temperature swings. Blossoms fall off the plants if nighttime temperatures are cooler than 60°F, and temperatures in the 40°F range can shock the plants so badly that they won't grow at all for a week or more. To minimize transplant shock, wait until nighttime temperatures stay consistently above 60°F to set out 6- to 8-week-old transplants.

Q: What's eating my peppers? The fruits have holes in them and when I cut them open, I found pale-colored worms.

A: The European corn borer probably has found your peppers. Despite its name, this pest isn't all that fussy about its diet and will feed on corn, peppers, beans, and other plants in the northern and central United States and southern Canada. The flesh-colored worms are the larvae of adult moths, which are 1 inch long and creamy, yellowish brown.

To combat the European corn borer, apply *Bacillus thuringiensis* (BT) and plant herbs such as dill, coriander, and parsley to attract braconid wasps and tachinid flies, which are the pest's natural predators. At the end of the season, pull up and destroy all pepper plants and spent cornstalks, where the pest overwinters.

The Great Pepper Cover-Up

Want to improve your pepper yields and protect the plants from heat and frost? Blanket your babies with a row cover or plastic! In cool climes, use floating row covers to keep your seedlings warm for the first few weeks after you plant them outdoors. In the South, drape row covers (or shade cloth) over a wood or PVC frame in midsummer to protect the plants from intense sun. And in any region, cover the plants with clear plastic on cold autumn nights to ripen the last few peppers.

GARDENER TO GARDENER

Sun Shades for Transplants

For their first few days in the garden, pepper (and other) transplants can use a little protection from the sun. To shade my tender seedlings, my husband made folding A-shaped frames from 1 × 2-inch cedar stakes. The frames measure 30 inches long and 18 inches wide when open. To adjust the height, you just loosen the bolt that holds the legs together, move them closer or farther apart, then retighten the bolt.

For the coverings, I use old, washable open-weave curtains cut to fit. I attached loops to the corners of each covering to hold it in place. When I'm not using the frames, I fold them up for storage and toss the coverings in the washer. (Note that, with plastic coverings, the frames make great season extenders, too.)

Betty Martin
Renton, Washington

Splint Pepper-Laden Branches

As a paramedic, I've been trained to use splints to help heal broken bones. Now I use the same technique in my garden. Whenever a fruit-laden branch from one of my pepper plants bends downward to the ground, I support it by taping two tongue depressors along the sides of the branch. The splint allows the fruits to mature naturally on the plant.

John Favicchio
Flushing, New York

Make Your Own Pepper Flakes

Have more peppers than you know what to do with? Make your own dried pepper flakes! It's easy and it really reduces storage space problems.

Pick your peppers, cut them open, and remove the seeds. (Remember to wear gloves when handling hot peppers.) Lay the cut pieces on a screen in a dry, airy location or place them in a food dehydrator. After the peppers dry, grind them up and jar them. Bell peppers make for a great sweet pepper shake. Hot peppers make a spicy one.

James J. Kelly
Clairton, Pennsylvania

Pepper Le Pew

Put your peppers to work as pest deterrents! We've found that dried hot peppers repel many animal pests. I grow the very hottest peppers I can find, dry them, don rubber gloves, then crumble the peppers over the plants and beds that I want to protect. One treatment lasts several weeks.

Judith Miller
Sandpoint, Idaho

Perennial Peppers

You can keep your pepper plants growing and producing year after year. I discovered this by accident when I dug up a couple of pepper plants that had been in my garden but hadn't yet produced ripe fruit. I potted the plants in large containers filled with a mix of garden loam and compost, then stuck them next to a bright window in a cool room. Over winter, I simply watered the pots whenever the soil felt dry, and the plants produced a few ripe peppers. In spring, I transplanted them back out to the garden. The plants are now 3 years old and several feet tall, and they produce a bumper crop of big, beautiful peppers.

John Daniels
Lansing, Michigan

Chapter 6

June

Cultivating Flowers Organically: Roses

The summer's short and ornament is what I want—all vividness.
Give me bright blossoms against the teeming green. Give me
orange flags, blue horns, white faces, yellow wings. Give me the
purple throat, breathless, of calla lilies—and red, red, red, red, red!

—*Andrew Hudgins*

With roses, gardenias, and mock oranges all at their peak now, it's little wonder that so many people choose to marry this month.

But you don't have to hold a wedding to enjoy June's lush beauty. Why not celebrate the season's beauty and bounty simply by having a garden party with friends? Nature has already provided the perfect party decor. Flowering perennials, bulbs, and shrubs are in full bloom. Gourmet treats—strawberries, asparagus, and peas—are close at hand in the food garden. And if you're lucky enough to have a spreading shade tree or an arbor draped with fragrant roses, you've also got a great place for guests to gather.

Unfortunately, many organic gardeners shy away from growing roses because they think roses can't be grown without toxic chemicals. Well, if you're one of those rose-reticent gardeners, here's good news: As you'll read on the coming pages, countless old-fashioned roses are resistant to the many diseases that plague modern roses. Most are deliciously fragrant and many re-bloom throughout the season. Plant a few of these terrific cultivars, and you'll have plenty of reason to celebrate outdoors, all summer long!

Gardener's To-Do List—June

**If you don't know what USDA hardiness zone you live in,
check the map on page 230 to find out.**

Zone 3

- ☐ Thin leafy greens and eat your thinnings.
- ☐ Harden-off and then set out melon and squash plants.
- ☐ Plant beans, cucumbers, and sweet corn.
- ☐ Get tomatoes and peppers into the ground by midmonth.
- ☐ Start brussels sprouts, cabbage, Asian cabbage, endive, and kale for fall.
- ☐ Brighten up beds with colorful warm-weather annuals such as marigolds, celosia, and impatiens.

Zone 4

- ☐ Plant the rest of your tomatoes, and set out peppers, eggplant, and basil.
- ☐ Sow more beans, sweet corn, leaf lettuce, and cucumbers.
- ☐ Midmonth, start seeds of cauliflower, broccoli, and brussels sprouts for the fall garden.
- ☐ Sow winter squash directly in the garden for a fall harvest.
- ☐ Near month's end, sow cilantro for the salsa you will make later this summer.
- ☐ Mulch blueberries and raspberries.
- ☐ Dress up drab corners with marigolds, zinnias, and other easy-care summer flowers.

Zone 5

- ☐ Mulch your potatoes.
- ☐ Trellis peas, and stake or cage your tomatoes.
- ☐ Make second plantings of bush beans and sweet corn.
- ☐ Sow pumpkin seeds for a Halloween harvest of jack-o-lanterns.
- ☐ Plant dahlias and gladioli bulbs, then stake them right away so that you don't damage the roots later.
- ☐ Midmonth, thin apples and peaches to 6 inches apart so the trees won't carry more fruits than the limbs can handle.
- ☐ Start seeds of brussels sprouts, broccoli, and cabbage for the fall garden.

Zone 6

- ☐ Stake or cage your tomatoes, mulch them, and remove any leaves that show signs of early blight.
- ☐ Set out sweet potato slips.
- ☐ Make second sowings of squash, cucumbers, sweet corn, and bush beans.
- ☐ When spring lettuce bolts, replace it with field peas, okra, or more tomatoes.

- [ ] When your peas are finished, replace them with pole beans, cucumbers, or asparagus beans.
- [ ] Pinch back chrysanthemums to make them bushy.
- [ ] Prune spring-flowering shrubs after they finish blooming.

Zone 7

- [ ] You still have time to plant heat-loving field peas, luffas, and asparagus beans.
- [ ] Start seeds (or root cuttings) of early-maturing tomatoes for a fall crop.
- [ ] Cut and dry bunches of thyme, oregano, and mint.
- [ ] Snake a drip irrigation line through your tomato patch before the plants begin to sprawl.
- [ ] Pull up cold-loving pansies when they succumb to the heat, and replace them with annual vinca, zinnias, or celosia.
- [ ] Water all of the fruit trees and bushes that you planted this spring.

Zone 8

- [ ] Harvest, weed, and mulch!
- [ ] Replace spring peas, lettuce, and potatoes with field peas, limas, or a summer cover crop of soybeans.
- [ ] Hand-pick Colorado potato beetles and other pests.

- [ ] Plant chrysanthemums, balsam, celosia, begonias, salvia, dusty miller, geraniums, vinca, and verbena.
- [ ] Harvest blackberries and blueberries.
- [ ] Mulch figs and muscadine grapes.

Zone 9

- [ ] Replace peas with beans, substitute Swiss chard for spinach, and put in taro after harvesting early potatoes.
- [ ] Start another generation of zinnias, sunflowers, and marigolds for late-summer bloom.
- [ ] Pinch back chrysanthemums.
- [ ] Water plants in the morning so they don't become susceptible to fungus and insect infestation.
- [ ] Solarize pests and diseases in vacant beds by wetting the soil, then covering it with clear plastic for about a month.

Zone 10

- [ ] On that rare cool day, create a new bed for fall greens. Cultivate the soil, add organic matter, wet the bed down, then cover it with clear plastic for at least a month to kill weed seeds and nematodes.
- [ ] Prune cassia trees, poinciana, bougainvillea, and jasmine after they bloom.
- [ ] Prune litchi, mangoes, and other tropical fruits after the harvest this month.

Until the late nineteenth century, almost every hybrid rose came from species that tolerated or resisted *black spot*, a fungal disease. These tolerant "heritage roses" (albas, gallicas, polyanthas, banksians, hybrid perpetuals, rugosas, etc.) produce blooms that are mostly scarlet, pink, or white.

Trouble was, everyone wanted a vivid yellow rose. Breeder Joseph Pernet-Ducher finally achieved that goal in 1893 by crossing a hybrid perpetual rose with a cultivar of *Rosa foetida*, a species that evolved in a climate where black spot does not thrive. The resulting golden orange seedling, 'Soleil D'Or', is the ancestor of most modern hybrid teas and floribundas whose flowers are yellow, orange, or fiery red. Unfortunately, the new colors came at a terrible price: susceptibility to black spot.

Roses without Chemicals

Although cultivated roses are descended from hardy, vigorous wild brambles, many gardeners believe that they are tender, little things that can't be grown without toxic sprays. Well, it's true that roses *can* suffer from diseases, but you can manage those diseases and keep your garden chemical free by using a two-pronged approach—select disease-resistant cultivars and handle any diseases that do show up with environmentally friendly methods.

Selecting Resistant Varieties

Let's start with disease resistance. Although many modern roses are highly susceptible to diseases (due to breeder emphasis on flower attributes alone), you still can choose from hundreds of widely available old-fashioned roses with outstanding disease resistance. (For descriptions of common rose diseases, see page 108.) When you order, tell the supplier you want varieties that can be grown organically in your area, and ask if your choices are suitable. Here are some favorites:

Hybrid musks. This class of highly fungus-resistant shrub and semiclimbing roses derives from complex crosses between *Rosa moschata* ("musk rose"), *R. multiflora* ("many-flowered rose"), and various cultivated roses. Hybrid musks bear their flowers in large clusters and bloom repeatedly throughout summer. They are the most shade-tolerant of roses and require little pruning.

Recommended cultivars include 'Belinda', which has single light pink flowers on a 7-foot shrub, and 'Ballerina', a 5-foot shrub bearing clusters of bright pink flowers with white eyes. 'Pax' and 'Prosperity' both bear clusters of white semi-double flowers on 7-foot shrubs.

'Buff Beauty', possibly the finest hybrid musk, forms a 7-foot shrub with clusters of full biscuit-colored blossoms that fade to white. Its fragrance is like a pineapple-banana fruit smoothie.

Hybrid rugosas. These shrub roses derive from the wild Asian seaside species *Rosa rugosa* ("wrinkly rose"). Their distinctly wrinkled leaves foil most fungal attacks except rust. Rugosas also are tolerant of salt and drought, and they're exceptionally frost hardy. Although rugosas naturally form large bushes, you can keep them in bounds with light pruning. They make great hedges and, once established, produce two or three rounds of fragrant flowers per year.

The best-known rugosas are 'F. J. Grootendorst' (clusters of dark pink double flowers that are serrated like carnations); 'Pink Grootendorst' (like 'F. J.' but medium pink); 'Fimbriata' (a white 'Grootendorst'); and 'Hansa' (a striking purple semi-double). Fragrant 'Sarah VanFleet' (medium pink) and 'Mary Manners' (white) also are excellent choices.

'Therese Bugnet', usually classed as a hybrid rugosa, is actually a complex cross between several sub-Arctic species and modern hybrids. This 6-foot-diameter shrub has striking gray-green foliage and bears large, double pink flowers with a light fragrance. Besides being disease resistant, it's exceptionally tolerant of cold, heat, wind, and alkaline soils.

Polyanthas. Although the polyantha class of small, repeat-blooming shrubs tends to be susceptible to powdery mildew disease, 'La Marne' is the exception to the rule. At 4 feet, 'La Marne' is larger than most polyanthas and resembles a hybrid musk. It bears erect trusses of deep pink semi-double flowers with white eyes, grows extremely well in light shade, and is highly resistant to fungal disease.

Teas. True tea roses, the ancestors of modern hybrid teas, are among the easiest roses to grow in hot climates such as Florida, parts of Texas and California, and the Gulf Coast. Compared to hybrid teas, the blossoms of true tea roses are a bit smaller and on less-erect stems. But the plants are very tolerant of heat and drought. Recommended cultivars include 'Duchesse de Brabant' (shell pink flowers); 'Madame Joseph Schwartz' (white with a blush center); and 'Mrs. B. R. Cant' (tightly packed red-pink blooms at their best in autumn).

Temperate-climate gardeners can grow disease-resistant hybrid teas, such as 'Radiance' (pink) and 'Etoile de Hollande' (red). Both are beautiful old standards and almost as trouble free as hybrid musk roses. Other easy-to-grow hybrid teas include 'Lady Diana' (a tall plant with long, elegant stems of peach pink blossoms); 'Macartney Rose' (highly fragrant red-pink blooms on a spreading shrub); 'Angela Lansbury' (heat tolerant, with shell pink flowers); 'Chris Evert' (orange-yellow roses from June through December); and 'Elina' (pale yellow blooms).

Austin "English roses." British hybridizer David Austin crossed modern cultivars with heritage roses to create this relatively new class of

---

You Decide

Although the following roses are highly *susceptible* to fungal disease, they are still sold for their pretty flowers. Avoid them, or, if you must grow them, try bordering the plants with large perennials to hide their oft-unattractive lower branches.

HYBRID PERPETUALS
'Geant des Batailles' (or 'Giant of Battles')
'Mrs. John Laing'

HYBRID TEAS
'Chrysler Imperial'
'Double Delight'
'Signora'
'Talisman'
'Whiskey Mac'

RAMBLERS
'Dorothy Perkins'

GRANDIFLORAS
'Gold Medal'
'Queen Elizabeth'

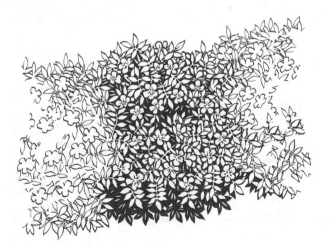

Rugosa roses can be planted as a hedge to create privacy and screen out unwanted views.

fragrant, old-fashioned-looking flowers. The plants rebloom as many times per season as hybrid teas and floribundas.

Because their parentage is so mixed and their introduction recent, long-term nationwide evaluations of disease resistance are not yet complete. But so far, several cultivars appear to offer superior disease resistance. West Coast gardeners have had good results with 'Graham Thomas' (up to 8 feet tall with fully double, golden yellow flowers); 'Belle Story' (semi-double, fragrant pink flowers on a 5-foot shrub); 'Othello' (double, reddish purple flowers on plants that do well in cooler areas); and 'Claire Rose' (an 8-foot shrub with fragrant, soft pink flowers). In the East, try 'Abraham Darby' (peach pink blooms on a tall shrub that can be trained as a climber).

Banksians. These semi-evergreen climbers, derived from the Chinese *Rosa banksiae*, cover themselves with hundreds or thousands of 1-inch flowers in very early spring. They bloom only on old wood, so only dead wood should be pruned.

Rx for Rose Diseases

Do your roses suffer from any of the following symptoms? Here's the probable cause and what you can do about it:

Symptoms	Cause	Solution
Leaves turn yellow and fall to the ground; remaining leaves develop black patches with fringed margins.	Black spot (fungus)	Clean up and discard all fallen leaves; water at ground level; use baking soda and oil sprays; plant resistant varieties.
Foliage has white, feltlike spores; leaves curl and turn purple; flower buds die without opening.	Powdery mildew (fungus)	Space plants far enough apart for good air circulation; mulch with compost; irrigate at ground level early in the morning; use baking soda spray; plant resistant varieties.
Plants are covered with tiny bright orange spores; leaves fall to the ground.	Rust (fungus)	Plant resistant varieties; use sulfur sprays.
Leaves have circular white dots with reddish margins (occurs mostly on glossy leaves of climbers).	Anthracnose (fungus)	Plant resistant varieties; use sulfur sprays.
A bright yellow pattern appears on leaves; plants are stunted and weak.	Mosaic (virus)	No cure. Buy plants from nurseries that use virus-free rootstocks and budwood, or that propagate root cuttings from virus-free mother plants.

Banksians are hardy only in Zones 8 through 11. But the plants are thornless, completely immune to rose diseases, unattractive to aphids and beetles, and can grow to an immense size. A 130-year-old *Rosa banksiae alba plena* (also known as 'White Lady Banks') that grew in Tombstone, Arizona, measured more than 65 feet in diameter! This species rose bears small, white, double flowers that smell like violets. If you live in the South, have lots of growing space, and are the kind of gardener who delights in raising things like giant pumpkins, this is definitely the rose for you. For something a little less vigorous, grow the sweetly fragrant, yellow 'Lady Banks' rose and allow it to scramble up a tree or large trellis.

Combating Fungal Diseases

Planting fungus-resistant rose varieties is the biggest step you can take toward disease-free organic rose growing. But if the bright colors of modern hybrid teas and floribundas tempt you, or if you've already planted such roses and now want to wean them from toxic sprays, don't despair. It's easy to treat even these "tender" plants the safe, organic way.

- Start with regular cultural cleanliness. Pick up and dispose of, *but do not compost*, infected leaves.

- Always water roses at their bases, never on their leaves.

- If you're already growing cultivars that are susceptible to disease, interplant them with resistant cultivars to greatly slow the spread of disease spores between plants.

- Go easy on nitrogen fertilizers—using too much can result in lush, sappy growth that attracts aphids, which transmit diseases as they feed.

- Mulch the bed to keep spores from splashing up on plants when it rains. A 1-inch-deep layer of compost can actually prevent disease.

- To improve air circulation, remove crowded canes and (in summer) remove the lowest leaves from the base of each plant.

- As a last resort, use natural sprays. For powdery mildew and black spot, apply a solution of 1 tablespoon each of baking soda and horticultural oil, with a few drops of insecticidal soap, per gallon of water once a week. Always test treatments on a small area first.

- For other fungal diseases, apply sulfur dust, wettable sulfur solution (2 heaping tablespoons per gallon of water), or Safer Garden Fungicide spray (a sulfur-based product) when the temperature is below 85°F.

- During the dormant season, spray with horticultural oil twice a month when the temperature is above 40°F.

What's Eating My Roses?

Although roses aren't bothered by many insects, they occasionally may be damaged by aphids, beetles, and mites. Here's how to keep them from ruining your display and compromising your plants' health:

Aphids (tiny, soft-bodied sucking insects that cluster on tips of buds or leaves): Dislodge with a strong spray of water or use insecticidal soap.

Japanese beetles (copper-colored adults that skeletonize leaves and chew holes in flowers): Hand-pick adults or spot-spray with insecticidal soap. Apply milky disease spores or beneficial nematodes to the lawn when grubs are in the soil.

Spider mites (microscopic pests that cause yellow, red, or gray patches on leaves; webbing sometimes appears on undersides of leaves): Spray insecticidal soap. Clean up the garden in fall. Apply dormant-oil spray in winter.

Rose Care: 1-2-3

Many organic gardeners shy away from roses because they think these classic flowers require hours of extra work. But the truth is, you can grow beautiful roses by investing just a few hours per *season*. Combine careful cultivar selection, proper planting, and a few basic maintenance techniques, and you'll be rewarded with gorgeous flowers and attractive landscape plants.

Proper Planting

If you buy dormant bareroot roses, plant them either in early spring or late fall. You can plant actively growing container roses anytime during the growing season, but the earlier the better.

When you're ready to plant, select a site that receives at least 6 hours of sunlight per day. (In very hot climates, however, roses can benefit from some midday shade.) Be sure the site has good air circulation. Avoid areas close to heat-reflective surfaces, such as concrete walks or south-facing brick walls, which can attract mites. The soil should be loose and crumbly, with excellent drainage. If yours isn't, work in lots of compost or create raised planting beds.

Give rose bushes plenty of room. A good rule of thumb is to plant shrub types (old-fashioned roses, floribundas, polyanthas, and hybrid teas) no closer than 3 feet apart. Plant climbers, ramblers, and large-growing shrub roses (such as many of the David Austin English roses) 6 to 8 feet apart. Space miniature roses about 1 foot apart.

Container roses. Planting container roses is easy. Dig a hole large enough to hold the container, making sure that the top of the soil in the container is just below ground level. Tip the container and tap it to release the soil, or cut the container away from the rootball with a utility knife. Set the rose in the hole, holding onto the top of the rootball rather than grasping the rose by its stem. Fill in around the rootball with soil. Water the rose and add more soil, if necessary.

Bareroot roses. To plant bareroot roses, choose a day when the soil is relatively dry. (If conditions aren't right for planting when the roses arrive, set the plants in a bucket of water until you're ready.) Prune off any damaged or dead roots, as well as dead or damaged canes, cutting back to a healthy bud.

Dig a hole large enough for the roots to spread out naturally, then make a mound of soil inside the planting hole. The mound should be high enough that the bud union is at the proper depth (usually 2 inches, but for exceptions see "How Deep to Plant Roses" on page 112.)

Set the rose on the mound, spreading out its roots. Completely cover the roots with soil, tamp gently, and water well. The soil will settle and fill the hole.

Easy Maintenance

After you've got your roses planted, maintaining them organically is a simple matter of mulching, watering, feeding, and pruning. Pay attention to these basics and you should prevent most rose disease and pest problems. If diseases or pests do show up, manage them organically using the methods described on the previous pages.

Mulching. Surround the plants with a mulch to conserve soil moisture and help keep roots cool through summer. Mulching also helps prevent black spot disease by keeping the fungal spores from splashing up onto the plants from the soil. Choose mulches that will decompose and add organic matter to the soil, such as compost, shredded leaves, or shredded bark.

Depending on the material you use, the mulch layer should be 2 or 3 inches thick. (A 1-inch layer of compost, however, is all that's needed to suppress diseases.) As the mulch breaks down over time, work it into the top few inches of the soil, then add fresh mulch to bring the depth back to 2 or 3 inches.

Watering. Rugosa roses are very tolerant of dry conditions, but most other species and cultivars shine with just a bit of extra watering beyond what Mother Nature provides. Once a week, water the plants

Set bareroot roses over a mound of soil in the planting hole and spread out the roots. Completely cover the roots with soil, tamp gently, and water well.

thoroughly—enough to soak down through their entire root area. Use a soaker hose or other form of drip irrigation, if possible. Overhead watering with a hose or sprinkler can promote the spread of diseases such as black spot, powdery mildew, and rust.

Feeding. Before you apply any fertilizers, it's a good idea to have your soil tested to determine any specific nutrient deficiencies. If your soil appears to be in good shape, put your roses on this simple diet regimen to keep them robust:

◆ About 2 weeks after spring pruning, apply a balanced organic fertilizer or a dressing of compost. (Don't feed newly planted roses until they've completed their first bloom cycle, though.)

◆ After roses have finished their first heavy bloom cycle in early summer, give them a second application of fertilizer.

◆ Feed reblooming roses again, after the second bloom cycle. (Don't use nitrogen fertilizers within 6 weeks of the first fall frost in your area, however. Late feedings encourage tender growth that frost will kill.)

◆ Always water your roses the day before feeding and again the day after applying fertilizer. This prevents fertilizer burn.

Pruning Basics

Don't be too nervous about pruning your roses incorrectly—they are very forgiving, even if you make a wrong cut or two. Each year, the plants produce strong new shoots, giving you beautiful blossoms and another try at pruning.

The best time to prune most roses is either late winter or early spring, about 4 to 6 weeks before the last killing frost in your area. Summer pruning is really just a matter of removing faded flowers

---

How Deep to Plant Roses

Most roses are grafted—one type of rose cultivar is joined to the rootstock of another. You can recognize these roses by the swollen or knobby area on the trunk, called the *bud union*. Most roses should be planted with the bud union 2 inches below the soil level to protect it from the cold.

The planting depth for hybrid teas and grandifloras, however, depends on how cold it gets where you live:

Minimum Winter Temperature	Placement of Bud Union
Above 32°F	1"–2" above surface
20°–32°F	Just above surface
Below 20°F	1"–2" below surface

throughout the growing season. Because pruning promotes soft new growth, discontinue it in late summer to avoid subjecting plants to winter damage.

What to Prune

The toughest part about pruning roses is deciding what to cut off and what to leave alone. As a rule of thumb, most vigorous roses can be pruned back to 6 inches tall, while others can be pruned to 10 to 12 inches without harm.

All roses. Remove canes that are dead, diseased, flowering poorly, crossed or rubbing, or growing inward. If suckers emerge from the area that's grafted, just below the ground, don't just cut them off at ground level or they'll resprout. Instead, follow each shoot underground to where it joins the stock, then use a trowel to snap it off.

Hybrid tea roses. Cut off the oldest stems at the base, leaving three to six healthy canes. Prune the remaining canes to shape the plant.

Floribunda roses. Remove the oldest stems at the base, leaving six to eight healthy canes. Prune off the top third of each remaining cane.

Climbers. Prune climbers to train and shape them, cutting back side branches to about 6 inches. Leave more canes than on other roses—about 15 or so—to get a good flush of bloom next year.

Ramblers. Ramblers bloom on first-year wood, so you should prune them more heavily each year. Cut them back to 12 to 18 inches from the ground each year.

Shrub roses. Prune off the top third of new canes and cut side shoots and older growth to 4 to 6 inches.

How to Prune

Pruning roses is a snap if you follow this advice:

- Make cuts ¼ inch above an outward-facing bud—the point where a leaf is or was attached to the stem.

- Cut at a 45-degree angle away from the bud. This promotes open growth in the center of the bush, encouraging air to circulate and preventing disease. Angling the cut prevents moisture from gathering on the bud and rotting it.

- Cut back to at least a five-leaflet node. Otherwise, the sprouting stem will be weak and spindly.

- If the inside, or pith, of the cane is discolored, cut the shoot back until the pith is white and healthy.

- If, after pruning, two or three shoots emerge from one bud, pinch off all of them but one.

Rose Companions

Roses complement many plants in the landscape. Choose these plants according to what you like, the color of your roses, and the character of the site. Here are some ideas to get you started:

- Ornamental grasses add a graceful linear quality.

- Blue-flowering plants complement the warm shades of red, pink, yellow, and peach.

- White-flowering plants provide a nice contrast to the deep-colored blooms of many roses.

- Plants that repeat the bloom colors of your roses are a good choice because they won't clash.

- Gray- or silver-leaved plants, such as lamb's-ears, catmint, and artemisia, provide a soft contrast.

- Low-growing plants can help cover the bare lower stems of rose canes.

Feeding Roses Naturally

Fertilizer needs vary based on your type of soil and growing conditions, but no matter where you live, your roses are going to need the essential nutrients.

A regular fertilizing schedule will help to keep your roses robust. Of all the plant nutrients, nitrogen (N), phosphorus (P), and potassium (K) are needed in the greatest quantities. These elements are represented in the above order in the three numbers listed on fertilizer bags and boxes.

Nitrogen. Plants need more nitrogen than any other nutrient. With adequate nitrogen, plants have strong, sturdy growth with lots of leaves and flowers. Nitrogen leaches rapidly from the soil and plants quickly use it, so you'll have to supply it throughout the growing season.

Phosphorus. Phosphorus is found in much smaller amounts than nitrogen in plants, but it's essential for root growth. Phosphorus also helps plant tissues to mature, which is critical for winter hardiness.

Potassium. Potassium helps plant foods move throughout the plant, and without it, roses will have poor flower production. If your garden soil is sandy, you'll need to apply potassium fertilizers more often.

Other nutrients are important, too, including sulfur, calcium, and magnesium. Your roses also need trace elements (called micronutrients) in smaller quantities.

Fertilizing Dos and Don'ts

- Do have your soil tested before adding fertilizer. Your county cooperative extension agent can help you obtain a test inexpensively.

- Do fertilize about 2 weeks after spring pruning.

- Do apply fertilizer after the heavy blooming period in early summer to encourage reblooming.

- Do fertilize again after the second bloom cycle. (If you live in Zone 9 or 10, you can fertilize your roses twice after the second bloom cycle, about 4 to 6 weeks apart.)

- Don't add nitrogen-rich fertilizers within 6 weeks of the first-frost date in your area. Late feedings encourage tender green growth, which frost (even light ones) will kill.

- Don't feed or fertilize newly transplanted roses until they have completed their first bloom cycle.

- Don't use as much fertilizer for miniature roses as is recommended for other rose varieties.

Top 10 Tips
for Long-Lasting Cut Roses

Bring the aromatic fragrance and timeless beauty of roses indoors. These tips will make your cuttings last longer.

1. Cut roses either in the evening or early in the morning, when the plants are most filled with water.

2. Gather rosebuds and half-open flowers, but avoid fully open roses, which will not last long.

3. Use pruning shears, not scissors, to cut roses.

4. Cut stems just above a five-leaflet leaf at a 45-degree angle.

5. Carry a bucket of water with you as you make your cuttings and plunge the stems into the bucket as soon as you've cut them.

6. Recut each stem underwater before placing it in a vase, cutting at a sharp angle so that the maximum area of the cut end is exposed to water.

7. Set the container of roses in a dark, cool, humid place (such as a basement) for at least several hours before using them in an arrangement.

8. Prolong the life of the roses by changing the water daily.

9. Remove all leaves that will be below water level.

10. Make your own conditioning product by mixing 1 teaspoon of vinegar, one aspirin tablet, and 1 tablespoon of sugar in 24 ounces of water.

GARDENER TO GARDENER

New Roses from Old

Starting new rose bushes from existing plants is easy, especially in late summer or early fall, when the wood has hardened off for the season. First, cut two stems with six sets of leaves each from the bush you want to propagate. (I always take two stems so that I have a backup if one dies.) Next, remove two sets of leaves from the bottom of each stem. The place where the leaves attach to the stem is called the node; roots will grow from these nodes.

Find a sunny spot in your garden, wet the soil, and push a thin piece of wood into the soil to create a hole. Now place the bare portion of one of the rose stems into the hole and tamp the soil down around it. Cover the cutting with a wide-mouthed glass jar to create a mini-greenhouse. Push the jar firmly into the soil to make sure it is secure. Leave the jar in place throughout winter.

The following spring, check the jar for signs of new growth on the stem. As soon as you spot buds, remove the jar. Care for the new plant just as you would your other roses, but don't fertilize it at all the first season. Expect at least one bloom the first year!

Sarah Rauch
Brooklyn, New York

Keep Out the Chill

Make insulating collars for roses from recycled 1-gallon plastic nursery pots. Use a utility knife to cut out the bottom and up one side of each pot, so that it opens up like the letter C. Wrap a pair of these cut pots around each rose bush, with the ends of the pots overlapping. Fill the pots with leaf mulch or compost. Remove the collars and insulation in spring.

Mary Leunissen
Guelph, Ontario

Natural Rose Trellis

Use a bare crabapple trunk to support the lanky stems of climbing or rambling roses. Plant a young rose 3 feet away from the tree, positioning the rose so that the stems lean toward the trunk. Build a framework of bamboo stakes to support and direct the stems up the trunk. When the rose canes reach the first tree branch, they'll be able to climb on their own.

Judith McKeon
Philadelphia, Pennsylvania

Rose Cones at Stake

Plastic rose cones are useful for protecting tender roses during cold winters, but I find that they blow away in strong winds, so I make my own anchors.

Using 1×1-inch lumber, I make T-shaped stakes. Each stake measures 8 or 9 inches long and has a 2-inch crossbar about $\frac{1}{2}$ inch from the top. To secure each cone, I hammer two stakes into ground on opposite sides until the crossbars reach the cone's flange.

Louis Bush
Pocatello, Idaho

Keepsakes for Kids

Here's an easy way for children to have fun and stay busy in the garden while you work. Buy a wide roll of masking tape and tear off two pieces about 6 inches long. Place one piece, sticky side out, around each of the child's wrists. Now your little helper can gather flower petals, blades of grass, small pebbles, etc., and stick them on the bracelets. After the bracelets are removed, put them in a safe place where the plants can dry and become little treasures of time spent in the garden.

Esther Brown
Danville, Kentucky

Chapter 7

July

High Summer: Tending the Garden

You want to make some honey?
All right. Here's the recipe.
Pour the juice of a thousand flowers
Through the sweet tooth of a bee.

—*X. J. Kennedy*

When summer's heat begins to drag on, you might decide it's time to head for the hills (or coast) to find some relief. July is a popular month for getting away from the responsibilities of work and home.

Trouble is, most gardens *never* take a summer vacation (unless you live in the very deep South, where all the rules are different!). If anything, gardens do just the opposite: In the steamy conditions of high summer, plants and insects step up their activity! Many vacationing gardeners come home to a jungle of overgrown weeds and pest-eaten produce. Or, if it's a drought year, they return to a barren wasteland instead of a garden.

Must gardeners forego their summer vacations? Absolutely not! If you care for your garden organically—using organic mulches, natural pest control techniques, and compost—your plants will be strong enough to manage quite nicely without you for a few weeks. Weeds and pests won't get out of hand, and the soil will contain enough moisture and nutrients to sustain healthy plant growth even during a few weeks of drought. With such an easygoing garden, you might decide to vacation in your own backyard this year.

Gardener's To-Do List—July

**If you don't know what USDA hardiness zone you live in,
check the map on page 230 to find out.**

Zone 3

- ☐ Continue to direct-sow bush beans, carrots, turnips, and beets.
- ☐ Start your first fall runs of lettuce, spinach, and other greens in the shade of taller crops.
- ☐ Mulch your potatoes!
- ☐ Gather and dry basil, mint, and other herbs before they flower.

Zone 4

- ☐ Set out seedlings of broccoli, brussels sprouts, cabbage, and cauliflower for fall harvest.
- ☐ Direct-sow some bush beans.
- ☐ Sow leafy greens such as mustard, Chinese cabbage, turnips, and lettuce, placing them where they'll get some afternoon shade.
- ☐ Plant fall spinach late this month.
- ☐ Sneak some potatoes into empty spots to harvest as "new" this fall.
- ☐ Deadhead annual flowers to keep them neat.
- ☐ Trim and repot gangly houseplants.

Zone 5

- ☐ Hurry and plant more bush beans, cucumbers, carrots, and summer squash.
- ☐ By midmonth, set out seedlings of cabbage-family members for fall harvest.
- ☐ Sow fall lettuce, Oriental greens, daikons, and snap, snow, or shell peas.
- ☐ Gather and dry herbs and everlasting flowers.
- ☐ Mulch your potatoes again.
- ☐ Set out asters and mums and sow some fast-growing annual flowers in the vacant spaces of flowerbeds.

Zone 6

- ☐ Direct-sow carrots, beets, and chard.
- ☐ Stretch the season with midmonth plantings of cucumbers, bush beans, and squash.
- ☐ By month's end, set out broccoli, cabbage, brussels sprouts, and cauliflower.
- ☐ Sow snap, snow, or shell peas.
- ☐ Plant any sprouted potatoes from your kitchen for a late crop of tender, new spuds.
- ☐ Water all fruits as needed through the harvest.
- ☐ Prune out old wood from raspberry and blackberry patches.

Zone 7

- ☐ Early in the month, sow a little more sweet corn and set out new tomato plants.

- ☐ Plant pumpkins and winter squash in a shady spot, but where the vines can soon run into the sunlight.

- ☐ During the second half of the month, start seedlings of broccoli, cabbage, cauliflower, and brussels sprouts indoors.

- ☐ Prepare beds for fall crops by sowing them now with a cover crop of fast-growing field peas or other legumes.

- ☐ Harvest blueberries and mulch strawberries heavily to protect them from heat and drought.

Zone 8

- ☐ You still have time to plant field peas, okra, limas, watermelon, and asparagus beans.

- ☐ Now's the time to solarize sick soil. Water empty beds, then cover them with clear plastic for at least a month.

- ☐ In the warmest parts of the zone, start tomato and pepper seedlings indoors.

- ☐ Set out chrysanthemums and pinch them back.

- ☐ Sow zinnias and sunflowers where you need late-summer color.

Zone 9

- ☐ Start tomato, pepper, and eggplant seedlings indoors for planting in fall.

- ☐ Outdoors, use old sheets or shade cloth to protect peppers and tomatoes from sunburn.

- ☐ Set out more sweet potato slips.

- ☐ In lower elevations, continue to plant corn, squash, beans, and field peas.

- ☐ In desert areas, prune back tomatoes by two-thirds.

- ☐ In humid regions, set out tropical vines and shrubs during rainy spells.

- ☐ Solarize empty beds that are infested with root-knot nematodes, or where diseases have been a problem.

Zone 10

- ☐ Direct-seed pole beans and limas, cantaloupes, collards, sweet corn, okra, southern peas, and watermelon.

- ☐ Indoors, start seeds of eggplant, peppers, and tomatoes now for the fall garden. Choose heat-tolerant cultivars, such as 'Heatwave'.

- ☐ Inspect passion fruit vines daily for caterpillars. When necessary, spray with *Bacillus thuringiensis* (BT), a natural caterpillar killer.

- ☐ Add everything you can to the compost pile. You'll need lots of humus to mix into the soil in fall.

Listen to Your Weeds!

What do you do when you see a weed in the garden? Jump in and frantically hack away with a hoe? Throw up your hands in despair? Learn something?

Yes, learn something! Those weeds are excellent indicators of soil conditions. In fact, experts known as geochemical botanists often look for specific weeds to help them locate minerals in the soil and to pinpoint geological features. You can apply this science in your own backyard in two ways: to plant garden crops that will thrive in the same conditions as those weeds or to amend your soil so that the conditions are less inviting to the weeds you find there.

Here are the most reliable weedy indicators and what they reveal about your soil.

Weeds That Say Your Soil Is Soggy

If you see **dock, foxtails, horsetail,** and **willows**, you can expect your site to suffer swampy conditions some time during the year. Other weeds that thrive in wet soil include **goldenrod, Joe-Pye weed, oxeye daisy, poison hemlock, rushes,** and **sedges**.

What could you possibly grow in such conditions? How about a fabulous garden filled with plants that like wet feet? Ornamental willows, including pussy willow and curly willow, will flourish here and provide plenty of material for flower arrangements. You can also grow dogwoods, Japanese iris, Siberian iris, yellow flag, ligularia, cardinal flower, and turtlehead. Don't grow invasive wet-loving plants like purple loosestrife or meadowsweet, however. They can overwhelm the area and destroy the natural balance of the wetlands.

Also, don't try to change these conditions. Wetlands are priceless natural habitats that are rapidly being lost to development. Besides, trying to "correct" such a site usually is a lost cause—in Nature, water almost always wins.

Weeds That Cry Out "Compaction and Crust"

Chicory and **bindweed** are telltale signs of compacted soil. That's why you often see the blue flowers of chicory along highways. Chicory also is common in gardens where beds have been left empty or, worse still, where soil has been worked when it's wet.

If your weeds indicate compacted soil, plant a cover crop of white lupines and sweet clover. They have roots as strong as those of pesky chicory, and they help to break up the soil. At the same time, these cover crops replenish the nitrogen levels in the soil.

Joe-Pye weed

Chicory

Although a hard crust on your soil's surface can prevent many vegetables and flowers from breaking through, it doesn't deter **quackgrass** or **mustard family** weeds at all.

If weedy mustard is flourishing in your garden, pull it up and plant closely related brassica crops such as broccoli, cabbage, cauliflower, and choy instead. They can push through crusty soil with ease. Replace quackgrass with a fast-growing grassy cover crop (such as rye) in fall, then till it under the following spring. The cover crop will loosen the soil and choke out the weeds.

To aerate and lighten crusty and compacted soil, add compost. Prevent future problems by working your soil only when it's dry.

Weeds That Signal Your Soil Is Sour

Dandelions, mullein, sorrel, stinging nettle, and **wild pansy** all thrive in "sour" acidic soil (pH below 7.0).

If you find these pests in your garden, grow plants that also like their soil on the tart side: hydrangeas (whose flowers achieve their most beautiful shades of blue in acidic soil), blueberries, rhododendrons, and azaleas. In the vegetable garden, endive, rhubarb, shallots, potatoes, and watermelon all tolerate soil with a pH as low as 5.0.

Or, if you'd rather grow plants that thrive in neutral soils, you could raise your soil's pH by applying dolomitic limestone. To determine how much lime to use, send a soil sample to a lab for testing, then follow the lab's recommendations. Wood ashes also will raise soil pH, but don't use any more than 25 pounds per 1,000 square feet, and avoid applying them more often than every 2 or 3 years. Compost is a better buffer: Just add enough to raise your soil's pH to 6.5 or 6.8.

Stinging nettle

Weeds That Say Your Soil Is Sweet

Campion, field peppergrass, nodding thistle, salad burnet, scarlet pimpernel, and stinkweed all indicate a "sweet" alkaline soil (pH higher than 7.0).

Ornamentals that do well in alkaline soil include lilacs, Persian candytuft, dianthus, baby's breath, helianthemum, dame's rocket, lavender, and mountain pinks. Some edibles also tolerate soil that's a little on the sweet side, including asparagus, broccoli, beets, muskmelons, lettuce, onions, and spinach.

If you want to lower the pH of your alkaline soil, add peat moss or elemental sulfur at a rate suggested by soil test results. Or, again, simply add compost regularly to bring the pH closer to neutral.

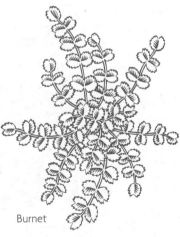

Burnet

Weeds That Warn of Worn-Out Soil

Biennial wormwood, common mullein, daisies, mugwort, wild carrot, wild parsnip, and **wild radish** are sure signs that your soil has poor fertility.

Luckily, many perennials actually flower better when the amount of food in the soil is on the lean side. This list includes achillea, antennaria, artemisia, asclepias, centranthus, cerastium, coreopsis, echinops, eryngium, gaillardia, salvia, santolina, solidago, and stachys. In the edible department, beans (and other legumes), beets, carrots, parsnips, peas, radishes, sage, and thyme tolerate soil that's low in fertility.

Of course, you could and should improve the fertility of at least some of that soil. First, have your soil tested. If the test reveals major deficiencies, correct them with organic fertilizers such as fish meal (for nitrogen), bonemeal (for phosphorus), and greensand (for potassium). From then on, use compost and cover crops to maintain your soil's fertility.

Weeds That Reveal Your Soil Is Rich

Fertile soil will often make its richness known by supporting happy and vigorous colonies of **chickweed, henbit,** and **lamb's-quarter**. In addition, a flush of **redroot pigweed** indicates an abundance of nitrogen in the soil, while **knapweed** and **red clover** reveal an excess of potassium. Spot lots of **purslane** and **mustard**? They could be telling you that your soil is rich in phosphorus.

To take full advantage of your soil's fertility, plant heavy feeders, such as corn, broccoli, lettuce, melons, squash, tomatoes, and peppers.

Wild carrot

Purslane

7 Secrets of Weed Warriors

Experience is often the best teacher, as every gardener who has ever let a weed go to seed knows. To find out the very best ways to prevent and eliminate garden weeds, the editors of *Organic Gardening* magazine asked readers for their most successful weed-beating strategies, tips, and techniques. Here are their top secrets:

1. *Timing is everything.* Don't think that you have to eradicate each and every weed the moment you see it. If weeds are too small, you may not be able to grasp them well enough to pull them out completely. On the other hand, waiting until weeds mature to control them isn't effective, either. By then, the wind may have carried their seeds throughout your garden, or their roots could have grown deeper than your cultivation tools will reach.

Each weed has a certain size that's most responsive to your tug. When the soil is slightly moist and friable and the weed reaches that ideal in-between size, give a pull—it should come out with little resistance and without snapping off.

2. *Hot water works.* Many gardeners say that pouring boiling water on weeds is the most foolproof eradication method. Most weeds are killed within a day of receiving such treatment. Thistle and other tenacious types, however, may need to be scalded several times before they'll surrender.

3. *A little sodium may do the trick.* To keep weeds from growing between the paving stones of walkways or patios, *Organic Gardening* readers often pour ordinary table salt (sodium chloride) in the cracks. Weeds rarely grow back, they report.

As an alternative to using salt, one reader sprinkled baking soda (sodium bicarbonate) directly on several weeds in his lawn. He found that the baking soda was more effective than hand pulling at killing such deep-rooted species as hairy cat's ears, dandelion, and broadleaf plantain.

After experimenting with various application methods and amounts, he came up with the following recommendations:

- Apply 1 teaspoon of baking soda per weed. Using less than a teaspoon won't kill the weed, but using more than that isn't necessary.

- Sprinkle the baking soda over the entire weed. Placing the whole dosage on the center of the stem of each plant is less effective than covering the leaf surface.

(continued)

Queen-Anne's-Lace: A Welcome Weed

Certain weeds, such as Queen-Anne's-lace, can actually benefit your garden. That's why some gardeners allow one or two of these attractive plants to remain in out-of-the-way areas of the garden. Like other members of the umbel family, Queen-Anne's-lace provides nectar for the beneficial insects that prey on pests. It also provides food for swallowtail butterfly caterpillars.

The trick is to learn to recognize these plants when they're small, so that you don't uproot them along with unwanted weeds. Take a look at the leaves: Queen-Anne's-lace seedlings have the same feathery foliage that characterizes other members of the family, including carrots and dill.

Mulch-by-Mulch Comparison

Covering your soil with a thick layer of mulch will block weeds, retain soil moisture, and feed the soil as it decomposes. But what's the best organic mulch for your specific situation? A quick guide:

Leaves. Best for earthworms. Shred and age them before using.

Grass clippings. Decompose fast. Best when dried before use.

Compost. A 1-inch layer prevents plant diseases and feeds plants.

Newspaper. Great for blocking weeds. Top with compost, grass clippings, or straw.

Bark. Most durable and attractive. Beware of treated wood painted to look like bark.

Pine needles. Attractive and abundant in some areas.

Straw. "Cushions" tomatoes against soil-dwelling diseases. May contain weed seeds.

◆ Apply a second dose, if necessary. Although a single application of baking soda killed about 80 percent of the cat's ears plants, a second application was needed to control the plantain.

◆ Apply the baking soda at any time during the growing season. It seems to be equally effective whether applied in wet spring weather or during a summer drought.

Note: If you live in a dry climate where the soil may be naturally high in salt, be careful about using salt or baking soda to control weeds. Too much sodium could kill not only your weeds but also your garden plants.

4. You can sprout them out . . . If removing weeds one by one is too big a task in your garden, try managing their life cycle to suit your plans. In Europe, some gardeners and farmers spray beer on newly tilled land to stimulate the germination of weed seeds in spring. Two weeks later, the farmers till again, turning the weeds into instant *green manure*.

5. . . . or crowd them out. Speaking of green manure, be sure to try cover cropping, a classic technique for controlling weeds organically. One *Organic Gardening* reader moved to a new home that had no mulch materials for blocking weeds, so she planted a mixture of soybeans and buckwheat in vacant garden beds. The cover crops grew beautifully, shading out all but the most stalwart weeds. And because the soil was moist beneath the crops' leafy canopy, the surviving weeds came out with ease. As a bonus, the cover crops enriched the soil when she turned them under.

Buckwheat is an especially appealing cover crop because it grows fast and matures in just 30 to 50 days. What's more, its white flowers attract pollinating bees to your garden. You might also consider one of the many types of vetch or clover, both of which help increase the soil's nitrogen content while they shade out weeds. Cover-crop seeds usually are sold at farm supply stores.

6. When in doubt, mulch. Mulch is a very effective way to control weeds, as many gardeners have known for years.

Newspaper mulch makes a particularly good weed barrier. Here's what to do. When your plants are at least 2 or 3 inches tall (or just after you have transplanted them to the garden), surround them with a layer of newspaper, two sheets thick, and cover the paper with grass clippings. The clippings will keep the soil moist, while the paper prevents light from reaching weed seeds waiting to germinate. The newspaper and clippings will even yield benefits for your garden long after the weeds cease to be a threat. By fall, the paper will decompose and you'll be able to turn everything into the soil.

If you'd rather leave those clippings right on the lawn to fertilize the grass, use a thicker layer of six to eight sheets of newspaper and anchor it in place with pine needles, bark chips, or another organic mulch.

7. *Action must follow planning.* Armed with the previous ideas, you might be tempted to lay back and contemplate which approach is best for your garden. Before you do, consider this caution sent by another reader: "Every year I plan to conquer the weeds in my garden, but that's about as far as I go. I'm sorry to say that my method of planning—without follow-through—has proven ineffective. I don't recommend it to amateurs."

> ### Salad "Dressing" in Flowerbeds
> Many perennials come up late enough in spring to give weeds a good head start! To hold back those competing weeds, try planting lettuce, spinach, or mesclun (mixed salad greens) all around the young perennials. By the time the perennials are large enough to take over, the greens will be ready to pull and enjoy in salads.
>
> Or, if you're planning to plant tender annuals among the perennials, start the season with an early crop of cool-loving greens where the annuals will grow. When the lettuce begins to bolt (set seed), the soil will be warm enough to plant the annuals.

Newspaper and grass clippings make an especially effective weed barrier. Spread the paper two sheets thick, then cover it with grass clippings or another organic mulch.

*Needle Those Sow
Bugs!*

*We live in a high-altitude
pine forest, and I had already
been using pine needles as
mulch. One spring I had a
severe problem with sow bugs
going after my seedlings. Out
of desperation, I broke up some
of the needles and laid them
on the ground around the
seedlings, forming a pine
needle triangle at the base of
each one. The sow bugs did not
cross the needle barrier, and
the new shoots remained
unharmed.*

*C. Reason
Ruidoso, New Mexico*

10 Fast Ways to Control Pests

Did insect pests get the best of you and your garden last year? Were you witness to overnight attacks on your veggie crop? Are you a bit worried that this year might bring a repeat performance? Never fear! Here are 10 timely, timesaving tips to help you take charge of your garden's bad bugs *now*:

1. Mix your signals. A confusing mix of sights and scents can help deter certain insect pests. So try to increase biodiversity and avoid monoculture by mixing plants from different families. Instead of planting long rows of a single crop, plant onions alongside broccoli, tomatoes with basil and chives, and peas with carrots. Better yet, interplant edibles with ornamentals. Add a few hot pepper plants to your flowerbeds, or edge your vegetable beds with low-growing annual flowers, such as alyssum and dwarf marigolds.

2. Attract an airborne defense squad. One of the best ways to short-circuit an onslaught of pests is to attract an airborne cavalry charge of beneficial insects. Many beneficials—including the small wasps that prey on pest caterpillars—will gratefully take advantage of the flat-topped floral landing platforms offered by members of the umbel family, which includes dill, Queen-Anne's-lace, parsley, and carrots. (You have to allow the parsley and carrot plants to overwinter and grow into their second year to get those umbrella-shaped flowers that beneficials find so attractive.) Other plants beloved by beneficials include sweet alyssum, all kinds of mints, and chamomile.

3. Negate nematodes. Marigolds can greatly reduce the damage caused by root-ravaging nematodes—those tiny soil-dwelling wormlike pests—but only if you use them correctly. For the best effect, grow a thick stand of marigolds as a cover crop for a season, then turn them under the soil. The next year, plant whatever you like in that area—nematdoes won't be around to cause trouble underground.

4. Grow your own decoy (#1). Try allowing a single weed to grow as a decoy among your cultivated crops. Decoy crops may attract pests and help to keep the bad guys away from your other crops. Striped blister beetles, for instance, seem to prefer redroot pigweed to tomato plants growing nearby. To keep the insects from moving to your tomatoes, check the pigweed each morning and shake off any beetles into a bucket of soapy water.

5. *Grow your own decoy (#2).* You can trap flea beetles in a similar manner using arugula, the spicy salad green. Pesky flea beetles—a voracious pest of eggplants, brassicas, and potatoes—will flock to the arugula first. Use a handheld vacuum to suck the beetles off the decoy plants before they can make their way to your main crops. You may have to repeat the vacuum cleaner escapade a few times each season to keep ahead of the invading flea beetle army.

6. *Grow your own decoy (#3).* Knowing that aphids are attracted to all things yellow, the staff of Ecology Action in Willits, California, have learned to plant yellow nasturtiums at the base of tomato plants to lure aphids away from the tomatoes. Monitor the nasturtiums closely, they urge. After the flowers have drawn in the aphids—and before the aphids reproduce—pull out the decoy plants and destroy their load of insects.

7. *Set up traps!* Earwigs, sow bugs, pill bugs, slugs, and snails all have one thing in common: They like to hide out in damp, shady places during the heat of the day. To take advantage of this trait, lure them with attractive "trap nests"—boards, pieces of paper, seashells, broken crockery, etc. Get out early every morning to check each lure, then dump the trapped critters into a bucket of soapy water.

8. *Pull back the mulch.* Organic mulches such as straw and leaves prevent weeds, maintain soil moisture, and improve soil quality. Unfortunately, under certain conditions they also can provide a home for insects that feed on tender young plants, such as slugs, sow bugs, and pill bugs. If these pests typically pose a problem in your garden, pull your mulch at least 2 inches away from the stems and stalks of transplants and young seedlings.

9. *Pull up those covers.* Sometimes the best way to head off insect trouble is to stretch some row covers over your crops. Besides keeping out pests, such as cucumber beetles, squash bugs, and cabbage maggots, row covers speed crop growth by trapping a blanket of warm air around new seedlings and established plants.

10. *Take out the apples and the trash . . .* Cleaning up garden debris may not be the flashiest method of controlling pests, but it is certainly one of the most effective and, by far, the easiest. By allowing insect larvae to overwinter in your garden and orchard, you are locking yourself into a cycle of repeated infestation. To break the cycle, promptly clean up all faded flowers, spent crops, and fallen fruit at the end of the season.

GARDENER TO GARDENER

Double-Duty Slug Barrier

I solved my slug problems and also devised a way to water my tomatoes more efficiently, both with the help of 5-gallon buckets. I remove the bottoms from the buckets, then sink them into holes in the garden, leaving about 4 inches showing above ground. Then I fill each bucket with soil and plant one tomato in it. The buckets direct all the water toward the root system, where it's needed. And slugs seem unwilling to scale the sides of the buckets to chew on the plants.

Dorothy Wyffels
Tillamook, Oregon

Remove Those Roots!
   Don't leave uprooted perennial weeds in your garden. Pick up all the pieces and toss them into a hot compost pile! Otherwise, the weeds may re-root and come back to haunt you—especially if you pull them just before a rainy period.

10 Easy Ways to Beat Weeds

A single weed can produce as many as 250,000 seeds. Though some seeds are viable for only a year, others can lie dormant for decades, just waiting for their chance to grow. Buried several inches deep, the lack of light keeps them from germinating. But bring weeds to the surface, and they'll germinate right along with your flower and vegetable seeds.

Even if you're diligent at hoeing and pulling weeds, more seeds arrive—by air, by water runoff, and in bird droppings. You may accidentally introduce weeds by bringing seeds in on your shoes, clothing, or equipment or in the soil surrounding the roots of container-grown stock.

If you had more weeds then seedlings last year or are already feeling defeated by the number of weeds choking out your favorite plants, don't worry! These surefire tips will help you keep down weed populations during the growing season:

1. Know your enemy. Before you can determine your best defense strategy against weeds, you need to know what you're up against. Some weeds, such as miner's lettuce, chickweed, purslane, and dozens of grasses, are shallow-rooted annuals. Others, such as dock, comfrey, thistles, and certain runner grasses, are deep-rooted perennials. The two types require different control methods. Arm yourself with a good field guide, then identify and inventory your weeds. After that, you can . . .

2. Assault annual weeds when it's dry. Wait for the weather to be hot and dry for several days, then attack young annual weeds with a rake, hoe, or trowel. That way, the drought-stressed weeds are sure to shrivel and die, even if your cultivation doesn't remove the entire root of the plant.

3. Give perennial weeds a shower. The long taproots of perennial weeds cannot be pulled out when the soil is dry. To remove these weeds, wait for wet soil—either from rainfall or from your hose. If the soil is wet and loose, even pesky thistles should come out with their roots intact—which means they won't grow back!

4. Comb that grass right out of your beds. If invasive grasses, such as Johnsongrass or bermudagrass, threaten your garden, use a pitchfork to "comb" your beds before you plant in spring, suggests an *Organic Gardening* reader. Work the soil until it's sufficiently loose for planting, then go over the entire area with a pitchfork, stabbing into the ground and levering it back toward the soil's surface. The tines of the fork will catch any buried grass roots, which you can then remove by hand. This technique has removed about 90 percent of the grasses from a reader's market garden in Texas.

5. Become a mulching maniac. Deprive weeds of the light they need by covering bare soil with a thick layer of grass clippings, shredded leaves, pine needles, or other organic mulch. Any survivors that do manage to penetrate the mulch usually are so weak that you can easily remove them by hand.

6. Cook 'em. If you've got a large-scale weed problem, bake the plants beneath a sheet of clear plastic. For best results, wet the soil before you cover it with the plastic. Leave the plastic in place for at least 3 weeks—ideally, when the weather is hot and sunny. This method is especially effective against cool-season weeds and annual grasses.

7. Let lettuce help your peas. Peas and other shallow-rooted crops can be damaged easily by cultivating the surrounding soil. That's why broad-leaved weeds can easily overtake them. So why not establish an edible, living mulch to fight the weeds *and* provide an extra early-season crop? Sow seeds of a fast-growing leaf lettuce thickly between young pea plants. The lettuce will outperform the weeds, and you can harvest the lettuce thinnings as you pick your peas.

8. Squash pigweed. If you're faced with a pugnacious patch of pigweed, fight back by planting a mixture of squash and buckwheat. The vigorous squash and quick-growing buckwheat will easily overtake the weeds. At the end of the season, harvest the squash, pull out the vines, and turn under the buckwheat. The buckwheat will add organic matter and nutrients to the soil for next year's crops.

9. Berry your weeds. Use strawberries to smother weeds! These perennial fruits spread by runners and are vigorous enough to overcome many weeds—even in light shade. In mild-winter areas, such as Zone 9, they'll grow (and hold off weeds) all year long. Try growing them as a groundcover beneath blueberries and roses.

10. Till 'em two times . . . In Maine's chilly Zone 5, organic market gardener Eliot Coleman uses a tiller to battle redroot pigweed, the seeds of which can remain viable in the soil for years. Coleman runs the tiller through his beds as early as possible in the spring to bring the weed seeds closer to soil's surface, where they can germinate. That's right: Coleman encourages the weed seeds to sprout! Then, a week or two later, he tills a second time to clear the area of the young weeds before he plants his vegetables.

Spring Ahead in Fall

Never allow weeds to set seed, or you'll face an even bigger weed problem next year. It pays to be especially vigilant in late summer to fall, when daylight-sensitive annual weeds mature and produce seeds before they are killed by frost. No matter how busy you are with the harvest, take the time to pull out those weeds *before* they set seed—or you'll be hoeing forever next spring.

Rules for Reading Weeds

Before you draw any conclusions about what your weeds say about your soil, consider the following.

- Look for large populations of the same weed rather than just a few individual plants.

- Look for more than one type of "indicator weed." Two or more weeds that like the same conditions are stronger evidence that your soil provides those conditions.

- Consider the health of the weeds. Robust plants are good indicators; weeds that look pale or weak don't tell you much about your soil.

- Weeds that keep coming back year after year are especially good indicators of soil conditions. Their environment must be hospitable for them to survive (or reseed) from one year to the next.

Outsmart Lawn Dandelions

If you see dandelions in your lawn, it probably means that the surface soil has become too acidic due to a loss of calcium through the removal of grass clippings. Under these conditions, dandelions get a foothold because their long taproot can reach into the subsoil for calcium, while the short roots of lawn grasses cannot. With the grasses too weak to crowd out the dandelions, the resourceful weeds thrive.

What to do? First, contact your cooperative extension office and find out how to have a soil test taken. (It usually costs less than $10.) Then, if the soil test confirms your suspicions, dust your lawn with lime according to the recommendations on the test. In milder cases you can use compost to replace the calcium. Also, set your lawnmower to cut the grass 3½ inches high (no lower), and allow the clippings to remain on your lawn. By taking these simple steps, you'll deprive dandelions of their competitive advantage.

Weeds Can Be Good Guys, Too

You may find it hard to believe, but having a few weeds here and there is actually good. Weeds can attract and provide shelter for beneficial insects (those that eat the pesky insects that damage plants). Weeds can help protect gardens and especially lawns from disease attacks, too. Because many diseases attack particular plant species, having more than one species of grass in the lawn ensures that your entire lawn won't succumb to a single disease. And some weeds offer other benefits: For example, dandelions are edible, and white clover provides nutrients for the soil.

Top 10 Ways to Conserve Water While Caring for Your Garden

Water is vital to the life of your garden and yard, yet gardeners often waste this precious resource. Whether your area is suffering from a drought or not, follow these guidelines to cut down on water usage:

1. Water deeply, making sure to soak the root zone rather than the whole yard.

2. When landscaping, choose native plants that require little or no water beyond what nature provides.

3. Plant groundcovers and shade trees to help keep your yard cool.

4. Mulch your garden regularly to lock in soil moisture.

5. Use watering methods such as drip irrigation or soaker hoses to reduce evaporation by directing water to plant roots.

6. Closely space plants in your raised vegetable garden, so there is less area to water (make sure to provide enough room for root development).

7. Follow the principles of xeriscaping, a water-saving garden design method used by many gardeners in more arid climates.

8. Limit the size of your lawn by adding a deck, patio, or walkway, which also adds backyard enjoyment.

9. Connect a rain barrel to a gutter on your home to create a ready source of water for your plants.

10. Recycle household water from drinking glasses and steaming or cooking vegetables and pasta to water plants.

Drip Irrigation: The Way to Water Wisely

When Mother Nature fails to provide all the water that your plants need to perform at their peak, you've got to step in and provide that essential 1 inch per week yourself. But what's the best way to water? Drip irrigation, for several reasons:

♦ It uses water efficiently, delivering it directly to plants' roots (unlike overhead sprinklers, which waste a lot of water). Almost no water is lost to surface runoff or evaporation.

♦ It helps prevent fungal diseases that find a hospitable environment by the soil-splashing action of sprinklers and hoses.

♦ It provides precise control, allowing you to maintain a steady level of soil moisture and to water a larger area (at the same time and at the same rate) than with hit-or-miss sprinklers.

♦ It more than makes up for the cost and effort involved in design and installation in water savings and increased plant growth.

♦ It delivers water directly to the plants you want to grow and less is wasted on weeds. And the soil surface between plants remains drier, which discourages weed seeds from sprouting.

Consider Your Options

Depending on your garden size and watering needs, you could get away with a setup as simple as a soaker hose, or you could take the plunge by investing in a full-feature professional kit. Most gardeners opt for something in between. Let's take a closer look at the possibilities.

Soaker hoses. Also known as a weeper hose or oozer hose, this simply is a length of porous hose, often made of recycled tires combined with polyethylene. Water soaks through the entire surface of the hose.

Soaker hoses generally are very durable. They aren't prone to freezing or cracking, and many brands resist the degrading effects of the sun's ultraviolet rays. Still, soaker hoses can come apart or develop leaks if the water pressure is too high—generally, above 10 pounds per square inch. If you remember to never open your faucet more than one-quarter turn, you probably won't have a problem. Or assure yourself of staying within this limit by inserting a pressure regulator disc in the female end of the hose.

Drip tubing with punch-in emitters. These drip kits feature lengths of ½-inch-diameter polyethylene hose in which you insert small plastic *emitters*. The emitters come in different shapes and can release water at various rates—usually from a half-gallon to 4 gallons per hour.

The main advantage of this type of system versus a soaker hose is that you can target the water so that it is delivered to very specific areas. For instance, if you have shrubs or trees or crops spaced widely apart, you can insert your emitters so that they drip only where the plants are located.

With emitters or a soaker hose, the drip rate can vary from one end of the line to the other—especially if the length exceeds 100 feet and/or your garden slopes significantly. The solution is a drip kit with pressure-compensating emitters, which are designed to put out water at the same rate regardless of the length of the hose or the slope of your yard.

Drip tubing with in-line emitters. These tubes are just like the punch-in type, except that the emitters are preinstalled inside the solid polyethylene tube. Typically, the emitters are spaced 12 inches apart all along the length of the hose, but 18-, 24-, and 36-inch spacing is also available.

Drip-irrigation kits. These drip systems come with everything you need—typically a soaker hose, a solid polyvinyl hose, a pressure regulator disc, end caps to close off each length of hose, and various plastic connectors. More-sophisticated systems may even include special filters and backflow preventers.

Prevent Winterkill of Roses by Watering *Now*

To prevent winter losses among your roses, be more conscientious about summer watering. In August, roses produce fewer flowers and begin to store energy in their roots for spring. Even though it doesn't look like the roses are growing, it's vital that they have enough water during this time so that the roots will be able to support growth next spring. A typical rose plant needs 2 or 3 gallons of water every 5 to 7 days.

Plan to water your roses the entire growing season—right up until the time the ground freezes.

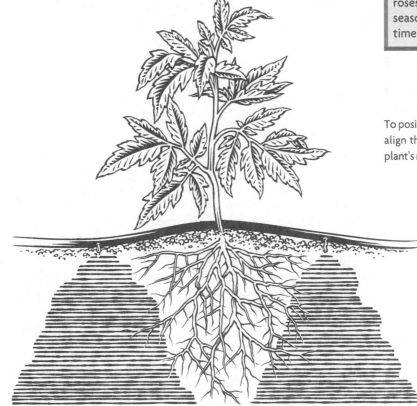

To position drip tubing with emitters, align the hoses so that you water a plant's roots, not the crown.

Soil Sample for Measuring Moisture

Deep watering encourages your lawn to send roots farther into the soil, so your lawn can last longer between rains or watering without becoming stressed. But how can you tell whether you're watering enough? The answer is underground. Start by watering for about an hour. After the water soaks in, use a spade to cut 3 to 5 inches into the soil. Then make a second cut about an inch away and lift the sample of soil out of the ground. You should be able to feel the point at which the water stopped penetrating. If it's less than 3 inches below the soil surface, you haven't watered long enough.

Putting It All Together

Don't go overboard by buying more watering equipment than you need. In the moderately rainy East, a simple soaker hose can do the trick. For a 5 × 40-foot ornamental border, for instance, two 50-foot lengths of hose should be adequate. Attach them to a single faucet with a twin coupler. When rainfall is scarce, turn on the faucet and let the hoses run for about an hour at a time. You can even let most types stay in place all season long, tucked beneath a couple of inches of mulch.

In drier climates, such as the interior West, you'll probably want a more complete system that can supply water to your entire property. To make the job easier, use electronic timers to turn the water on and off routinely. As the season progresses—and watering needs change—you can program the timer to keep the water running for longer or shorter periods. But if you use a timer, be sure to keep an eye on the batteries. If they run down, you may not know there's a problem until your plants start to wither.

No matter where you live or what type of drip system you use, the key to success lies in the proper placement of the tubing. Aim to create a zone of continuous moisture below the surface of your growing beds by placing emitters 6 to 18 inches away from the stems of your plants. This will spread moisture broadly below the soil surface and promote a very wide, deep root system. And by placing the drip tubing so that it doesn't drench the very center of the plant's root system, you avoid problems such as root rot.

Another advantage of creating this zone of moisture is that you don't have to worry so much if one or two of the emitters clogs because the overall moisture present in that area of the garden will compensate for the loss of the emitters.

How far apart should you space emitters to create such a moisture zone? For loamy soils, space them 18 inches apart; for sandy soils, keep the emitters 12 inches apart; and for heavy clay soils (which hold water naturally), space emitters 24 inches apart.

Finally, there's the aesthetic question: to bury or not to bury the hoses? Packing your hoses beneath the soil line may look better, but it can cause some problems. Tunneling gophers may be attracted to the sound of the running water and leave you with a munched-through line that you won't notice until your plants begin to fail. Emitters also can clog with soil as a result of the reverse flow that occurs when the system is shut off.

The answer is to compromise. Simply cover the tubing with 1 or 2 inches of mulch. This hides the hoses and assures that your drip lines are easily retrievable at any time for repair or modification.

GARDENER TO GARDENER

Garden Coaster

Looking for an easy way to transport garden tools, bedding plants, and pulled weeds? Try a snow coaster—the round metal disc (sometimes called a snow saucer) that kids use to slide down snowy slopes. In summer, I pull my coaster along with me as I weed, plant, and prune, piling up weeds and grass for recycling. It glides easily along behind me on my way to the compost pile. A coaster is much lighter and easier to maneuver than a wheelbarrow and has no wheels to damage the lawn. It's a great way to use an inexpensive toy year-round.

Sue De Kelver
Marinette, Wisconsin

Pine Bark Blocks Weeds and Slugs

Last spring, I accidentally discovered an excellent slug deterrent. We had purchased some pine bark nuggets on sale but found that the pieces were too big to use around seedlings. So we borrowed a friend's chipper/shredder and put the nuggets through it. The result was a superfine bark mulch that we used around our perennial seedlings. Not only did this attractive mulch block weeds and retain moisture, but it seemed to repel slugs. Everything else in our garden was chewed except for the seedlings surrounded by the pine bark mulch!

Barbara Libby
Columbia, Maine

Gum Ball Mulch Stops Slugs

Our huge sweet gum tree drops many thorny, round seed balls onto the lawn each year. We've also had a problem with slugs eating our hostas. We solved both problems by mulching around our hostas with the gum balls! The slugs dislike climbing over the thorny things, so our hostas now remain free from damage.

Ellen Newman
Webster Groves, Missouri

A Neighborly Affair

I am always in need of more mulch and compost, so I distribute a flyer to nearby homes to let people know that they can drop off their bagged leaves and herbicide-free grass clippings in our driveway. In my flyer, I point out that the air is polluted by burning leaves and that composting them is a better option. My neighbors respond by bringing me dozens of bagfuls. It's a win-win situation for everyone.

Hannelore "Honey" Huisman
Rock Island, Illinois

Runners Make Sturdy Ties

I've discovered that an abundant supply of free, readily available plant ties is growing right in my garden—strawberry runners. The runners are nice and long and remain intact even through long, cold winters. I use them to tie up roses, delphiniums, columbines, and pea vines. If heavy rains knock over my peony flowers, I pull off a runner and use it to encircle the stems and pull the flowers upright again. The color of the runners blends into the garden more naturally than string, especially after winter months weather the runners to a thin brown strand.

Janice Plante
Wasilla, Alaska

GARDENER TO GARDENER

3-Season Grasshopper Strategy

There may be some wild prairie magic about the sound of grasshoppers humming in the tall grasses, but for this gardener, a hopper's ability to decimate a garden is not enchanting. To keep them in check, I use a three-season strategy that really works:

1. Spring. *I mow down any stands of tall grass near the garden so grasshoppers don't have a place to congregate. When tilling the garden, I interrupt their life cycle by digging up their egg cases. I also use a trowel to dig around garden borders, next to the house foundation, and near the frames of raised beds—wherever egg cases can be found. (The cases look like slender clumps of rice grains.)*

2. Summer. *I keep the grass around my garden cut short. A neighbor boy helps me hand-pick any hoppers that venture in. (He uses the hoppers as fishing bait.) We collect in the morning when the pests are still sluggish, drown some of them in a can of soapy water, and toss the rest onto a platform bird feeder.*

3. Fall. *I turn over the edges of the garden after the first hard frost to expose as many egg cases to the killing cold as possible.*

Jeanne Cross
Lincoln, Nebraska

Pick(ling) Weeds

In our hot, humid region, weeds spread like wildfire. When I needed to prevent the weeds from becoming a jungle around our fence and sidewalk, I tried my grandmother's old-fashioned recipe—vinegar. I simply poured some vinegar on the weeds and they were dead within a day. An easy tug was all that was needed to pull them out of the ground.

Diane Merkle
Houston, Texas

Distract Asparagus Beetles

I tend a large asparagus patch, from which I've harvested many nice spears over the years—despite my constant battle with asparagus beetles. My patch is far too big for me to pick all the beetles by hand, so I was delighted to find an alternative, organic control method. After seeing lots of beetles on the tall ferny stalks of mature asparagus, I began to deliberately allow one spear to grow to maturity as quickly as possible, every 3 or 4 feet throughout the patch. Now beetles congregate on these tall stalks and hardly ever bother the emerging spears.

Ronald Spangler
Halifax, Pennsylvania

Chapter 8

August

Orchard Picks: Apples and Native American Fruits

Walnuts and pears you plant for your heirs.

—*Mid-17th-century proverb*

In many gardens, August brings a sort of intermission. The first plantings of peas, leaf lettuce, and brassicas have been pulled and replaced with fall crops that only now are taking hold. And except for ever-blooming marigolds, impatiens, and other annuals, the color in many flowerbeds has turned to solid green, as showy spring and early-summer perennials have faded for another year.

But if your property includes fruit trees or bushes, August is anything but quiet! Many fruits, including early apples, peaches, pears, and brambles, take center stage, providing baskets of healthful, sweet treats.

If you "inherited" an ancient apple or other fruit tree when you bought your property, you may have wondered who planted it, or how many families have enjoyed its fruit in sauces, pies, and cobblers. Planting fruit trees and shrubs is an act of faith—often, gardeners don't begin to enjoy "the fruits of their labor" for years. But once those fruits begin bearing, the rewards keep coming, sometimes for generations.

Planting and nurturing fruit is a long-term investment, so why not invest your time in fruits that are ideally suited to organic growing? With native fruit trees and shrubs, you (and the garden caretakers who follow you) can spend less time managing problems and more time enjoying the harvest.

Gardener's To-Do List—August

**If you don't know what USDA hardiness zone you live in,
check the map on page 230 to find out.**

Zone 3

- ☐ Finish planting your lettuce and spinach.
- ☐ At midmonth, pick off all tomato flowers so the plants' energy can be devoted to ripening fruits.
- ☐ Be ready to cover your tomatoes, peppers, and beans to protect against the first frost.
- ☐ Harvest everlasting flowers after a period of warm, dry weather.
- ☐ Renew mulches around roses, but stop feeding them.
- ☐ Prune out the oldest raspberry canes after harvest.
- ☐ Set out new strawberry plants.

Zone 4

- ☐ Water the garden as needed to keep everything growing fast.
- ☐ Make a last sowing of spinach and cold-hardy lettuce.
- ☐ Dig compost into beds before setting out Chinese cabbage and fall brassicas.
- ☐ Root cuttings of perennials and herbs to grow indoors this winter.
- ☐ Dig and divide irises and daylilies.
- ☐ Sow Iceland poppies and set out pansies for fall color.
- ☐ Protect grapes and late-bearing tree fruits from birds.

Zone 5

- ☐ Sow spinach, turnips, and cold-hardy lettuce for fall.
- ☐ Water the garden if it doesn't receive at least 1 inch of rain per week.
- ☐ If you want to create a new garden bed or rejuvenate an old one for next year, cultivate the space well and sow a cover crop of oats, rye, or ryegrass.
- ☐ Sow hardy biennial flowers, such as sweet William and forget-me-nots.
- ☐ Trim ragged tops of perennials that have finished blooming.
- ☐ Prune out old raspberry canes.
- ☐ Top-dress strawberries with compost.

Zone 6

- ☐ Plant a rainbow of lettuces, multihued radishes, and colorful kales for your fall salad garden.
- ☐ Plant plenty of spinach!
- ☐ Harvest potatoes, bulb onions, beans, and squash.
- ☐ Thin fall salad greens.
- ☐ Late this month, set out pansies for fall and spring.

- [ ] Sow Shirley poppies, larkspur, snapdragons, and sweet William.
- [ ] Collect seeds of favorite annual flowers.
- [ ] Water strawberries and transplant robust new runners.

Zone 7

- [ ] Early this month, start seeds indoors for the fall garden, including broccoli, cauliflower, cabbage, Chinese cabbage, and scallions.
- [ ] Sow carrots, beets, kale, and chard in the partial shade of taller plants.
- [ ] Make your last sowings of squash and cucumbers.
- [ ] Plant fall peas by midmonth, and set out brassica seedlings for fall harvest.
- [ ] When Labor Day is near, direct-seed kohlrabi, kale, and collards.
- [ ] Keep watering fall-bearing raspberries and everbearing strawberries; replenish their mulch as needed.

Zone 8

- [ ] Start some basil, cucumbers, and squash indoors and set the plants out a week after they sprout.
- [ ] Direct-seed bush beans, pumpkins, and sweet corn by midmonth, followed by peas and dill.
- [ ] Start carrots in a bed of loose soil.
- [ ] Prune back okra by one-third to encourage side shoots to bear.

- [ ] Start celery, broccoli, cabbage, and brussels sprouts indoors, then set out the seedlings when they have six leaves.
- [ ] Rake the orchard floor clean to interrupt the life cycles of pests.

Zone 9

- [ ] Sow watermelon early in the month, along with limas, southern peas, and sweet corn.
- [ ] After midmonth, begin setting out your tomato and pepper transplants.
- [ ] Direct-seed cucumbers, squash, and bush beans in partial shade.
- [ ] Plant basil and dill.
- [ ] Renew fading flowerbeds by filling them with fast-growing marigolds, zinnias, and annual vinca.
- [ ] Harvest tree fruits and use spoiled fruits as bait to lure green fruit beetles into narrow-necked jar or bottle traps.

Zone 10

- [ ] Set out tomato and pepper transplants after midmonth, then cover them with shade cloth to shield them from the summer sun.
- [ ] Start seeds of hardy perennials indoors, including pansies, gaillardia, and daisies.
- [ ] Direct-sow foxglove, larkspur, cosmos, and hollyhocks.
- [ ] Pinch chrysanthemums to encourage new blossoms.

Caring for Native Fruits

Native plants, including native fruit trees, are used to fending for themselves. In the wild, they thrive with no care. Likewise, when planted in your home landscape, they won't demand your attention the way other plants do. If you provide these locals with the soil and light conditions they prefer, insect pests and diseases rarely will be a problem.

During the first season, give native fruit trees the same care you would other young plants attempting to establish themselves—namely, keep weeds at bay and be sure to water during dry spells. Also cover the trees with netting for the first year or two to protect the ripening fruits from animals.

All-American Tree Fruits

Scattered throughout the woodlands of North America are hidden treasures waiting to be discovered—pawpaws, mulberries, and persimmons. These tasty native American fruits grow on beautiful trees that practically take care of themselves. What could be better for an organic garden?

American Persimmon

Over the years, our American persimmon (*Diospyros virginiana*) has received a bad rap, mostly due to the awful sensation that comes from eating the unripe fruits. But when ripe, the fruits actually have a rich, honeylike flavor and jellylike texture.

Native from Connecticut to Florida and west to Kansas, persimmon trees grow about 50 feet tall and look handsome in the home landscape. For home gardens in most regions, the best cultivars are 'Early Golden', 'Florence', 'Garretson', 'Killen', 'Morris Burton', and 'Wabash'. If you live in the fruits' northern range, however, choose an early-ripening variety, such as 'Meader', 'John Rick', or 'Yates'.

Persimmons are attractive trees with large, leathery leaves that turn beautiful bright colors in the fall. The bright orange fruit often hangs on the branches long after the leaves drop. Persimmon fruit can be very astringent before the fruit is mushy-ripe, but some cultivars can be enjoyed while still firm.

Persimmon pollination can be a little tricky to understand. Sometimes the male trees produce female flowers and vice versa. The easiest way to get fruits is to plant a self-fertile female tree, such as 'Garretson' or 'Meader'. To get fruit from the other varieties, you'll need to plant both sexes, or graft a male branch onto a female tree.

Dig deeply when you plant or transplant—the persimmon tree has a long taproot. Potted trees can be transplanted at any time, but the best time to plant a bareroot tree is in spring. Be sure to water the plants throughout their first season. Persimmons produce a lot of root suckers. Discourage them by spreading a thick layer of organic mulch such as compost over the root zone. Remove suckers whenever you see them.

Create a lightweight fruit picker with PVC pipe. Use a pipe that's wide enough for the fruits you will be harvesting to fit through. Cut a notch in one end of the pipe to hook the fruit. The snagged fruit will fall through the pipe into your hand or harvest basket.

Persimmons are reasonably pest-free in the home garden. They can be troubled by scale and borers, and by persimmon psylla and citrus mealybug in the South.

Don't harvest your persimmons until the fruits are fully colored and soft. The ripe fruits are delicious when eaten fresh, but they also make good pies, breads, cookies, and cakes. Before you use persimmons in a recipe, add ½ teaspoon of baking soda per cup of pulp to remove any remaining astringency.

Juneberry

Maybe you've already planted one of these 25- to 30-foot trees for its white or reddish spring blossoms and vibrant autumn foliage. If so, don't overlook the tasty, edible fruit. The small blue, red, or white berries have a unique sweet flavor that hints of almond.

Juneberry trees and shrubs (*Amelanchier* spp.) grow wild throughout North America and are known by various other names, including saskatoon, shadblow, and sarvis or serviceberry. All species bear edible fruits, but the tastiest ones are found on the Allegheny serviceberry (*A. laevis*), the thicket serviceberry (*A. canadensis*), the saskatoon (*A. alnifolia*), and a hybrid, the apple serviceberry (*A. × grandiflora*). Good varieties for fruit and beauty include 'Ballerina', 'Cumulus', and 'Robin Hill'.

Plant your juneberry in well-drained soil in either full sun or partial shade. The trees are hardy in Zones 4 through 8, and need little care once they are established. If you see a few orange spots on the leaves, don't be alarmed. It's probably rust, a disease spread from wild red cedars. You don't need to take any special measures because the disease usually doesn't harm the fruits.

Juneberries begin to bear fruits in their third or fourth year. Harvest them quickly—before they drop, dry up, or are eaten by the birds. You can eat them right off the tree or cook them, complementing their sweetness with one of the season's tarter fruits, such as currants. For traditional American fare, cook juneberries with rhubarb, or pound the dried berries with meat (preferably buffalo) to make pemmican, a staple of the Native American tribes of the Prairies.

Pawpaw

Want to grow a "tropical" fruit in a temperate climate? Try a pawpaw (or "Hoosier banana," as it is sometimes called), the northernmost member of the custard apple family and cousin to the cherimoya and soursop. The 10- to 25-foot trees (*Asimina triloba*) are native to woodlands from New York to Georgia and west to Nebraska, but their lush, drooping leaves give them an exotic appeal. The smooth, creamy

GARDENER TO GARDENER

Pucker Up for Persimmons

When harvesting wild persimmons, the first and most important rule is to wait until after a frost so that the fruits' flavor will be more mellow. Spread a piece of plastic beneath the tree and shake it. Gather the fruits that fall easily, then come back in a few days to harvest more. Like apples, persimmons ripen gradually over several weeks.

Carry the fruits home and put the slightly underripe ones on your kitchen counter. Wash the rest gently under running water, then sieve them or run them through a food mill. Use the fresh pulp in a favorite recipe or freeze it for later use.

Ruth Jacobs
Gallipolis, Ohio

GARDENER TO GARDENER

Cardboard Cats Foil Birds

My 25-foot-tall 'Hachiya' persimmon tree is very productive, but it's hard to keep birds away from the ripening fruits. Netting works okay for smaller trees but isn't practical for larger ones like mine. My solution? Scare the birds away with two 18-inch-tall cat decoys. To make the decoys, I simply drew the cats on a piece of cardboard box, cut them out with scissors, then suspended them from the tree.

Dan Arriola
Hacienda Heights,
California

fruits taste something like a banana with hints of mango and pineapple. They often weigh as much as a pound apiece.

You can grow pawpaws if you live within Zones 5 through 8. They are small, deciduous trees with purple flowers. The flowers aren't very prominent, but they do appear late enough in spring to escape frosts. Plant the trees in well-drained soil and be sure to dig a hole deep enough to accommodate their long taproots. You'll probably get the most fruits from a tree planted in full sun, but pawpaws are woodland natives, so light shade is okay, too, especially in the first year or two. Be patient—the trees grow slowly at first.

For the best fruits, plant 'Overleese', 'Mitchell', 'Taytwo', or 'Sunflower', advises the PawPaw Foundation, which promotes research and development of pawpaw cultivars. Plant at least two different varieties to ensure good fruiting. As further insurance, you could hand-pollinate the flowers with an artist's brush: Dust the brown pollen from one plant's flowers onto the shiny green, ripe stigmas of another. (In the wild, beetles and flies pollinate pawpaw flowers, but you may not have enough of these natural pollinators around to do the job for you.)

Pawpaws taste best if harvested when their yellowish skins become speckled with brown, a sign of full ripeness. For milder flavor, pick slightly underripe fruits and allow them to finish ripening indoors at room temperature. Delay ripening by storing them in the refrigerator before setting them out to ripen. Each mature tree will yield about 25 to 50 pounds of fruits a season.

The "proper" way to eat a pawpaw is to cut it in half and scoop out the flesh with a spoon, but the fruit is just as tasty if you peel back the skin and eat it like a banana. If you decide to cook your pawpaws, don't go overboard on sweetener or you'll steal their naturally good flavor.

Mayhaw

The name says it all: Mayhaw (*Crataegus aestivalis, C. opaca,* and *C. rufula)* is a southern edible hawthorn that ripens in May. The fruits vary in color from yellow to bright red, range in size from ⅓ to ⅔ inch across, and taste similar to a tart crabapple.

The broad-topped thorny trees are an attractive addition to the landscape—especially in early spring when they burst into a cloud of white to pale pink blossoms. They grow about 25 feet high and should be spaced at least 20 feet apart.

Plant at least two mayhaws to ensure cross-pollination and maximum fruit production. The trees grow and produce best when planted in well-drained, slightly acidic soil in Zones 6 to 9. (Although the trees are hardy to –15°F, they don't fruit well in Zones 5 and colder.)

In most regions, you won't need to do much to maintain an established mayhaw tree other than harvest it. The easiest way to do that is to spread a sheet under the tree and shake it. The fruits typically ripen over a period of several weeks. Expect a 5-year-old tree to yield about 5 gallons of fruits.

You can pop the mayhaws right into your mouth—fresh off the tree—but most people prefer to cook the fruits into marmalades, preserves, and desserts. Old-timers from the Deep South claim that mayhaws make the best jelly in the world. If you want to try your hand at mayhaw jelly, expect to get about 6 pints of jelly for every 6 quarts of fruit. As with pawpaws, go easy on the sweetener, though, or you'll mask the distinctive mayhaw flavor.

Mulberry

The mulberry's fruits look like blackberries but range in color from deep black to red to lavender to pure white. Their flavor ranges from strictly sweet to tangy sweet. Many "wild" mulberries actually aren't fully native, but are "half-breeds"—the result of a cross between our native red mulberry (*Morus rubra*) and the Chinese white mulberry (*M. alba*), which was introduced in the nineteenth century for the silkworm industry. Although some female trees need a male pollinator, many cultivars set fruit without pollination.

If given full sun and well-drained soil, mulberry trees are practically carefree. Provide plenty of growing space, however, because mature mulberry trees can reach 30 feet or more in both height and spread. Also try to find a site *away* from walkways and driveways so that the ripe fruits don't stain them or your shoes!

Plant container-grown stock any time the ground isn't frozen, or set out bareroot plants in spring or fall, while they are dormant. Space plants 10 to 30 feet apart. Mulberries are easy to care for: no need to prune, and you can handle dieback by simply cutting off infected portions.

Birds like the red mulberry's acidic red fruits, but you're more likely to enjoy the sweeter hybrids. The most widely available cultivar, 'Illinois Everbearing', grows well throughout most of North America and bears large, tasty, nearly seedless fruits throughout the summer. And like most mulberry varieties, it needs no cross-pollination. Although birds may eat lots of your mulberries, mature trees generally produce enough fruit for you and the birds.

Mulberries ripen over the course of several weeks. To harvest mulberries in quantity, spread a clean sheet under the tree and shake the branches. Ripe fruits do not keep well fresh, but they can be dried. For cooking, pick the fruit when it is slightly underripe.

Want to Know More?

A great way to learn more about growing fruit—including natives and other unusual types—is to join the North American Fruit Explorers (NAFEX). The organization publishes a quarterly magazine in which members share information. Write to NAFEX, 1716 Apples Road, Chapin, IL 62628; or visit www.NAFEX.org.

The Berries of America!

Long before North American gardeners grew cultivated varieties of European fruits, native berries sustained early settlers through cold winter months. European settlers brought to America their fondness for fruit, but many of the European plants adapted poorly to their new surroundings. As these imports succumbed to mildews and other maladies, determined fruit growers looked to the wild for hardy disease- and pest-resistant substitutes and found that the robust flavor of these untamed natives more than compensated for their lack of continental culture.

Today, these same fruits—either those in the wild or their lightly cultivated cousins—are your best bet for better berries.

Bramble Bonanza

Few fruits are as easy to grow and as tolerant of neglect as the members of the bramble tribe (*Rubus* spp.), which include native blackberries and raspberries.

No matter where you live or what your soil type, there's bound to be a bramble near you. More than 400 native species—many as tasty and productive as the "improved" garden hybrids—are found throughout North America. Blackberries alone include more than 200 identified species, and most of them will thrive just about anywhere. Raspberries are a bit fussier, requiring a fair amount of water for peak performance.

Blackberries fall into two principal categories: upright and trailing. The trailing types, also known as dewberries, ripen earlier and are less winter hardy than the upright types. Although some native blackberries are less suited than others for fresh eating, all can be used to make excellent wines, juices, jams, and jellies. For your home garden, you can choose named cultivars such as 'Darrow' and 'Illini Hardy', which combine the delicious wild berry flavor with improved hardiness and productivity. 'Navaho', a blackberry variety with thornless canes, is hardy to –20°F.

To succeed with blackberries, choose a site with good drainage. Southern dewberries will do fine even on dry, rocky soils not tolerated by other brambles. Trellis the plants so that air circulates around the foliage and fruits and discourage fungal diseases; as extra protection, avoid overhead watering by using a soaker hose or other drip irrigation.

Our wild American red raspberry (*R. strigosus*), common in many northern states, produces excellent fruits all summer long. Look for it along roadsides, in abandoned fields, and around the edges of woodlands, growing on bristly upright canes 2 or 3 feet tall.

Three other natives are the blackcap raspberry (*R. occidentalis*), a black raspberry found on burned-over land and in other disturbed areas;

the western thimbleberry (*R. parviflorus*), which grows wild in the cool hills of the upper Midwest; and the salmonberry (*R. spectabilis*), a hardy but not always pleasant-tasting raspberry of the Pacific Coast.

For the best crops of native or cultivated raspberries, make sure the plants receive plenty of water in midsummer, but not too much late in summer. Mulch the base of the plants with shredded leaves, pine needles, or straw to boost yields. Both raspberries and blackberries spread rapidly, so mow around them but take care not to cut their shallow roots.

Ribes *Jamboree*

Have a damp, sunny spot you don't know what to do with? It could be just the place to put in a carefree crop of currants or gooseberries. At least 80 species of this genus (known as *Ribes*) grow along creek banks, lake shores, and moist wooded valleys throughout North America.

The flavor of some of these tart native berries closely rivals their commonly cultivated European counterparts. But forget fresh—these fruits are better in jams, jellies, and pies.

The buffalo currant (*R. odoratum*) of the Midwest is probably the most familiar native currant. Also known as the Missouri or clove currant, the plant grows as tall as 10 feet and bears fragrant yellow blossoms and tasty black berries. Harvest is easy, as the branches are thornless and the fruits are borne in clusters. 'Crandall' is the most commonly available garden variety.

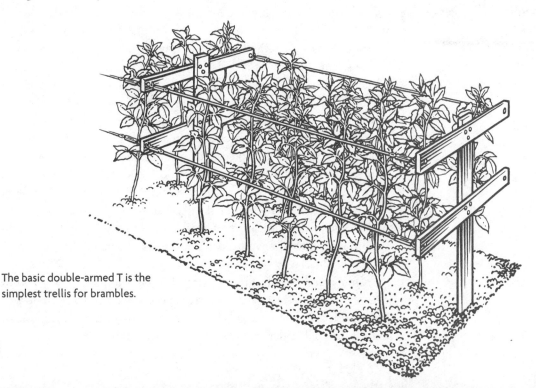

The basic double-armed T is the simplest trellis for brambles.

The *Ribes*-Rust Connection

Gooseberries and currants (*Ribes* spp.) are somewhat notorious in North America as alternate hosts of white pine blister rust. The disease requires the presence of both a susceptible *Ribes* and pine trees. Although white pine blister rust rarely kills *Ribes*, it can be devastating to pines. Because of this, the federal government restricted the growth of *Ribes* in the 1920s. Since then, research has clearly shown that only certain blight-susceptible currants pose a threat, and the federal ban was lifted in 1966. Some states, however, still have bans so check with your cooperative extension office before you plant.

Our native gooseberry species (*R. cynosbasti, R. hirtellum,* and others) produce tangy fruits that are excellent for pie fillings or preserves. On the downside, gooseberry branches have barbs that can make harvesting a challenge. But those who have acquired a taste for gooseberries believe the effort is worth the reward. To safely harvest gooseberries, hold a branch up with one gloved hand while you strip the berries with your other, ungloved hand. Improved cultivars, such as 'Welcome', 'Pixwell', and 'Poorman', are compact ornamental bushes that have little trouble with diseases and insects.

Keep harvested berries out of direct sunlight. Store fresh gooseberries at near-freezing temperatures and high humidity for no longer than 2 to 4 weeks. Currants don't store quite as long in the refrigerator as gooseberries do, so use them within a week or two.

As with most native fruits, gooseberries and currants need very little care once they're established. Provide full sun or light shade and loam soil with a pH between 5.5 and 7.0. Fertilize with plenty of compost, and mulch with 2 or 3 inches of organic material to keep soil cool and moist. By pruning out older branches (those more than ¾ inch thick) each year, you'll harvest larger berries.

If powdery mildew or leaf spot diseases are problems where you live, plant a disease-resistant cultivar.

Respectable Elders

The rich, full flavor of the American elderberry (*Sambucus canadensis*) makes it one of the favorite berries for use in wines and jellies. Throughout history, the berries and flowers have been used for teas, muffins, and fritters; as a pest repellent; and as a veritable cure-all. Although the medicinal value may be overstated, the juicy berries are one of Nature's most potent sources of vitamin C and also contain plenty of vitamin A, niacin, calcium, and thiamine.

The attractive shrubs reach about 6 feet tall and are easy to grow in full sun or partial shade. You can grow it in Zones 2 to 9. Plant two varieties for cross-pollination; popular garden varieties include 'Johns' and 'Adams'. If possible, provide rich, moist, but well-drained soil with a pH of 6.0 to 7.0. Elderberries don't like dry feet, so be sure to water the new plants regularly for at least the first year. For top yields, mulch around the bushes and fertilize annually with compost. Plants yield better crops if another elderberry is growing nearby.

One caution: Those lovely white flowers should give way to full clusters of purple-black fruit, *not red*. Species that produce red berries are considered poisonous and should be avoided—with the exception of the Pacific "red" elder (*S. callicarpa*).

Red and Blueberries

Both cranberries and blueberries belong to the genus *Vaccinium*. All of America's commercially cultivated cranberries come from a single native species, *V. macrocarpon*. Two other natives—both low, creeping evergreen shrubs—produce smaller fruits that are equally good for relishes and sauce. One is the small cranberry (*V. oxycoccos*), also known as bogwort, marshwort, mossberry, bogberry, and sourberry. It bears red fruits that are often speckled with white. The lingonberry (*V. vitis-idaea*), also called cowberry, partridgeberry, and mountain or lowbush cranberry, makes a handsome evergreen groundcover, with white to pink flowers in spring followed by tart, red berries in late summer. Both species are native to the northern United States and Canada.

Most cranberries need moist, acidic soil (pH 4.0 to 5.0), although lingonberries will tolerate drier soils. If you don't already have a boggy site on your property, amend your soil with peat moss or sand before planting. Mulch all types of cranberries with pine needles or wood chips in summer to increase productivity.

The cranberries' close cousins, the blueberries, may be America's favorite fruit family. These abundant berries are within the reach of gardeners everywhere, with native species and hybrids found in all regions. No matter where you live, you'll need to cover bushes with netting to prevent birds from picking the plants bare. Keep harvested berries refrigerated for no longer than 2 weeks.

The highbush blueberry (*V. corymbosum*) does well just about anywhere east of the Mississippi and in the Northwest, and is native to most of the eastern United States as far south as northern Florida.

The lowbush blueberry (*V. angustifolium*) is native from the Alleghenies to New England and Canada and widespread into the upper Midwest, where it is a valuable commercial crop. Only 6 to 18 inches tall, its low, spreading habit and attractive red fall foliage make it excellent in home landscapes.

The mountain blueberry (*V. membranaceum*), native to the Pacific Northwest, is valued for its exceptional flavor. Cultivated varieties have been selected from the native species and will grow well outside the original growing range.

Blueberries thrive naturally in soils with a pH ranging from 4.0 to 5.5. That's acidic with a capital A! If your soil pH is higher than that, add plenty of peat moss before planting, then mulch with 3 inches of an acidic organic mulch, such as pine needles, leaves, sawdust, or wood chips. Plant blueberries during spring in cold climes and during fall in warmer regions.

Berry Confused

The word "huckleberry" is often used interchangeably with blueberry, but true huckleberries belong in a different genus (*Gaylussacia*). Huckleberries resemble blueberries and have similar growing requirements. The main difference is that huckleberry fruits contain crunchy seeds, while blueberry seeds are soft and insignificant.

Give Apples a Strong Start

Yes, you *can* grow apples organically! If you select disease-resistant varieties (see "Organic Apple Picks" on this page) and handle your trees with care, problems should be minimal. When insect pests or diseases do show up, use organic controls such as those suggested in "Apple Pest Problem Solver." Here's how to get your apples off to a strong start:

1. Choose the best rootstock. First, be sure you have the best rootstock for your needs—dwarf, semidwarf, or standard. Each rootstock has its own advantages and disadvantages. For instance, trees with certain rootstocks may require staking, be less cold hardy, or have little resistance to fireblight and other diseases. Ask a local nursery, apple orchardist, or your county extension service to recommend the best rootstocks for your site and climate.

2. Plant it right. Plant your apple trees in either early spring or fall. (Avoid fall planting if you live in a very cold region.) Choose a well-drained site that receives plenty of sunlight. If your soil is poor, amend the entire site—not just the planting hole—with compost and other organic materials, preferably the season *before* planting.

Dig a generous-sized planting hole. Place some of the amended soil back in the hole so that the graft will be at least 6 inches above ground level. If you plant deeper, the graft could sink below the surface when the soil settles. Backfill the hole, then water the tree thoroughly.

When choosing rootstocks for your apple trees, consider your soil type, harvest needs, and available growing space.

Dwarf apple trees grow 7' to 10'

Semidwarf apple trees grow 14' to 17'

Standard apple trees grow 20' or taller

Trees on dwarf rootstocks need staking to compensate for the less vigorous roots. After planting, drive a sturdy wooden or metal stake into the ground near the tree. Tie the tree loosely to the stake. Wrap the trunk to protect it from sunscald, and install a hardware cloth sleeve to protect the trunk from rodents. Spread a layer of organic mulch over the area.

3. Mulch matters.

Apply an organic mulch, such as straw, after planting an apple tree. Mulch helps the soil stay moist and blocks the growth of competing weeds and grasses, which could stunt the tree's growth. Keep the mulch about 2 feet away from the trunk of the tree, however, and remove the mulch in fall so that rodents don't move in for the winter.

4. Feed lightly.

About 3 weeks after spring planting, spread a balanced organic fertilizer in a 2-foot radius around the base of the tree. Each year, spread 5 to 10 pounds of compost out to the drip line in late winter. Shoots on a well-nourished, young apple tree should grow 1 or 2 feet in a season; those on a mature, bearing tree should grow 6 to 10 inches. If growth is less than that, give them an inch or so of compost the following spring. If your tree fails to thrive, have a leaf analysis done through your extension service to identify any nutrient deficiencies.

5. Train a leader.

Train the tree to a *central leader*—the main trunk—so that the interior of the growing tree has access to plenty of sunlight and air.

Year 1: After planting, prune the trunk back to 2 or 2½ feet. Also head back any side branches by one-third to one-half their length.

During the growing season, when the new growth reaches 1 to 1½ feet, choose three or four main branches spaced about 4 to 8 inches apart along the trunk. Cut off all other branches on the tree, but leave the central leader alone.

Year 2: The next winter or early spring, remove any new shoots that are growing at a narrow angle and competing with the central leader. Also remove any vertical side branches. Head back each branch tip by one-third to one-half to encourage side branches to form.

When the new growth is 1 or 1½ feet long, choose another "layer" of three or four main branches about 1½ feet above the previous branches and cut off any other new branches. The remaining branches should create a spiral pattern up the trunk so that each has access to the sun.

Year 3 and after: Repeat the pruning cuts from Year 2. Each year, focus your efforts on choosing the newest layer of main branches.

Apple Pest Problem Solver

Insect pests taking a bite out of your apple harvest? Here's how to manage them organically:

Mites, aphids, and scale. If you observed infestations of these pests last year, apply dormant-oil spray very early in the season, when a quarter-inch of green shows on the flower buds.

Codling moths. Reduce codling moth numbers by wrapping trunks with corrugated cardboard to attract larvae looking for a site to pupate. Remove and destroy the wraps one month after each generation of larvae begin moving down the trunk. When apples are about 2 inches across, slip small brown paper bags over them and staple shut. Spray several applications of Surround, a kaolin clay product, beginning at petal fall.

Apple maggots. Hang red, sticky ball traps from mid-June until harvest (one trap per dwarf tree and six per standard tree).

Plum curculios. At petal fall, spray Surround. Or, reduce damage by shaking trees daily for several weeks, knocking beetles onto a tarp, then drowning them in soapy water. Clean up all dropped fruits in June and again in fall.

Top **10** Fruit Pests

Need a quick lesson on who's competing with you for your fruits and berries? This quick primer will help you identify and deter the most common fruit pests.

1. **Apple maggot:** The larvae of these pests leave apples with dimples outside and brown tunnels inside. Hang red sticky ball traps in affected apple trees.

2. **Codling moth:** Codling moth larvae chew their way into apple cores and destroy the fruit. Fight back by applying dormant oil sprays and using sticky tree bands.

3. **Fruit borers:** Blackberry, currant, gooseberry, and raspberry canes are weakened and often break due to the larvae of this pest. Prune affected canes.

4. **Leafhopper:** The leafhopper attacks most fruits and vegetables, especially apple trees and grape vines. Apply dormant oil sprays to kill overwintering adults, and wash nymphs from plants with a strong spray of water or insecticidal soap.

5. **Mealybugs:** Knock mealybugs from plants with a strong spray of water or insecticidal soap.

6. **Plum curculio:** These insects feed on fruit trees and blueberry bushes. Knock beetles onto a drop cloth and then destroy them.

7. **Rose chafer:** This reddish brown beetle is especially damaging to blackberry, grape, raspberry, and strawberry plants. In cases of severe infestation, drench soil with insect parasitic nematodes to kill larvae.

8. **Spider mites:** Spider mites suck juice from the undersides of leaves, weakening trees and reducing harvests. Rinse plants daily with water or insecticidal spray to rid them of the telltale webs most species spin on leaves and shoots. Dormant oil spray is also helpful.

9. **Strawberry root weevil:** Adult weevils feed on leaves and fruits, but the most severe damage is done by the larvae, which bore into the crowns and roots of plants, leaving wounds through which diseases then enter.

10. **Thrips:** Affecting a wide variety of fruits and vegetables, these minute insects scar developing fruit and stunt and distort plants. Spray dormant oil on fruit trees and attract native predators, such as pirate bugs, lady beetles, and lacewings.

Fantastic Ways to Enjoy Your Fruitful Bounty

Make the most of your fruit harvest with these two easy and oh-so-delicious recipes. Your hard work has never tasted this good!

Raspberries with Wild Rice

1½ cups defatted chicken stock

¾ cup wild rice

2 teaspoons canola oil

⅛ teaspoon dried marjoram

1 cup raspberries

2 tablespoons minced parsley

In a 1-quart saucepan, bring the stock to a boil. Transfer to the top of a double broiler. Add the wild rice, oil, and marjoram. Cover and cook over boiling water for 1 to 1½ hours, or until the liquid has been absorbed and the rice is tender. Transfer to a serving bowl. Fold in the raspberries and parsley.

Makes 4 servings.

Apples and Oats Breakfast Cereal

4 cups water

1⅓ cups oat bran

½ cup raisins

1 apple, shredded

1 tablespoon maple syrup

½ teaspoon ground caraway seeds

½ teaspoon ground cinnamon

1 to 2 cups skim milk

In a 2-quart saucepan, bring the water and oat bran to a vigorous boil, stirring constantly. Reduce the heat to low and cook for 2 minutes, stirring frequently, until thick. Remove from the heat and stir in the raisins, apples, maple syrup, caraway seeds, and cinnamon. Let stand for 5 minutes. Spoon into bowls and pour the milk over the cereal to serve.

Makes 4 servings.

GARDENER TO GARDENER

Net More Berries

My 'Black Satin' blackberries are luscious in July and August, but the Japanese beetles seem to love them almost as much as we do. For the past few years, I have thwarted these beetle attacks by using an inexpensive nylon net barrier to protect my precious berries.

As soon as the beetles begin appearing, I drape the bushes with the netting—the same kind used to make prom dresses and tutus. Using 72-inch-wide netting, I sewed together two widths, then fastened the netting over the branches with clothespins. The barrier easily covers the brambles and is tough enough to withstand strong winds much better than standard row-cover fabric.

> Laurie Maxwell Tenney
> Bluemont, Virginia

Best-Ever Cider Press

Making homemade apple cider is a great way to use up all those imperfect apples at the end of the season. But I've found that the commercial hand-crank grinders sold with cider presses don't really grind the apples finely enough to squeeze out all the juice.

So I bought a new garbage disposal unit and mounted it on a simple stand made from 2 × 4-inch lumber. When making cider, I set a 5-gallon bucket (lined with a large, nylon strainer bag) beneath the disposal to receive the apple slurry. When the bucket is half full, I simply lift out the strainer, leaving the cider in the bucket. With this method, I can quickly crank out nearly a gallon of pure juice into the bucket.

> Donald Yellman
> Great Falls, Virginia

Blooms Boost Yields

We've increased our raspberry yields by planting flowers near our berry patch. The flowers attract pollinating bees, which then move on to the raspberry blossoms. Last season, we picked 600 pints of raspberries that we sold to a local food co-op!

> Margaret and Edward Wanserski
> Rosholt, Wisconsin

Birds and Videotape

If birds are a bother, scare them away with the tape from an old videocassette. Remove the tape and cut it into 2- or 3-foot lengths. Hang the tape from the branches of the trees and shrubs. The slightest breeze causes the strips to wave and the movement keeps birds away.

> E. H. Berenson
> Chesterland, Ohio

Easy Berry Picking

Here's how to recycle plastic milk jugs into handy "buckets" that make picking berries a breeze:

1. Start with a clean, gallon-size plastic milk jug. Cut out a large opening that leaves the top and the two sides adjacent to the handle intact. Remove the other two sides to a point about halfway down the jug. Make sure the opening that you cut out is wide enough for you to comfortably drop in the berries you're picking.

2. To use the berry bucket, slide a belt through the handle and fasten it around your waist. Have another container ready to pour the berries into as the jug becomes full.

> Carol Kelly
> Saltsburg, Pennsylvania

Chapter 9

September

Bringing in the Bounty: Harvest and Stocking Up

Autumn, the bringer of fruit, has poured out her riches, and soon sluggish winter returns.

—Horace (65–8 B.C.)

There's excitement in the air this month. Just as spring brings a rush of activity to start the gardening season, the first month of autumn signals the approaching end of the growing season in most regions. For gardeners, cool nights and shorter days seem to trigger an ancient instinct to gather and preserve the garden's bounty in preparation for the dormant season ahead.

There's much to do. Tender vegetables and fruits soon must be harvested (or covered); garlic and spring-blooming bulbs need to be planted; and produce must be frozen, canned, dried, or pickled.

To save time and celebrate the season, why not hold a harvest party? Invite family and friends to bring some of their harvest, then work together to "put up" big batches of sauce, pickles, and frozen veggies and fruits. Afterward, enjoy the last cookout of the year, followed by a dance beneath the stars and harvest moon.

Gardener's To-Do List—September

**If you don't know what USDA hardiness zone you live in,
check the map on page 230 to find out.**

Zone 3

- ☐ Before the nights get too cold, pick all of your mature green tomatoes and store them indoors to ripen off the vine.
- ☐ Harvest squash, beans, and other tender vegetables.
- ☐ Cook up big batches of tomato sauce for freezing or canning.
- ☐ On freezing nights, cover lettuce, cauliflower, broccoli, and carrots to stretch the season.
- ☐ Wait 'til after frost to harvest and savor kale.
- ☐ Plant tulips, daffodils, and other spring-flowering bulbs.
- ☐ Clean and repair tools before storing them for winter.

Zone 4

- ☐ Extend the season! Plant salad greens, such as spinach, winter lettuce, and kale, in a coldframe.
- ☐ Before the first freeze, harvest tender vegetables, such as tomatoes, peppers, and melons.
- ☐ Can or freeze the last harvests of sweet corn, tomatoes, and beans.
- ☐ Prepare to cover broccoli, cabbage, and cauliflower.
- ☐ Toward month's end, dig up potatoes, onions, turnips, and carrots.

- ☐ Gather leaves and pine needles to use for winter mulching.
- ☐ Plant spring bulbs.

Zone 5

- ☐ Plant a fall salad garden of winter lettuce, spinach, and mâche in a coldframe or plastic tunnel.
- ☐ Remove the bottom leaves from brussels spouts plants to direct energy to the sprouts.
- ☐ Harvest winter squash, pumpkins, tomatoes, and peppers before frost.
- ☐ Pick apples, pears, and other late fruits, then freeze or can them to enjoy this winter.
- ☐ Prune herbs and geraniums, pot them up, then set them on sunny windowsills indoors.
- ☐ Plant pansies and spring-flowering bulbs.
- ☐ Don't cut back ornamental grasses; enjoy their feathery foliage throughout winter.

Zone 6

- ☐ Save seed from your best plants of heirloom beans, tomatoes, squash, and melons.
- ☐ Gather up and compost all of your spent garden plants.
- ☐ Harvest broccoli, cabbage, cauliflower, and kohlrabi if they're ready; if not, prepare to cover them on freezing nights.

- ☐ Harvest apples, then rake the orchard floor to disrupt the life cycles of pests.
- ☐ Weed blueberries and raspberries, then mulch them with chopped leaves.
- ☐ Plant pansies, spring-flowering bulbs, and hardy perennials such as daylilies.
- ☐ Sow a fall cover crop in vacant vegetable beds.
- ☐ Clean and repair tools before storing them for winter.

Zone 7

- ☐ Continue planting spinach, lettuce, radishes, arugula, Oriental greens, kale, and collards.
- ☐ Dig up sweet potatoes and peanuts while the weather is still warm; cure them before storing.
- ☐ Late this month, plant next year's garlic crop.
- ☐ Divide multiplier onions.
- ☐ Set out new strawberry plants.
- ☐ Start pansies, snapdragons, and sweet William from seed.
- ☐ Reseed and fertilize thin areas of fescue lawns.

Zone 8

- ☐ Set out brussels sprouts, kale, broccoli, cauliflower, and cabbage transplants; check them daily for leaf-eating pests.
- ☐ Direct-seed carrots, beets, lettuce, parsley, spinach, turnips, leeks, and kohlrabi; shade the beds until seeds germinate.

- ☐ Late in the month, sow sweet William, pansies, bachelor's buttons, poppies, and snapdragons.
- ☐ Plant trees and shrubs, then water them well weekly.
- ☐ Work compost into beds; replenish mulches.

Zone 9

- ☐ If scorching temperatures have eased up, set out tomato transplants by midmonth.
- ☐ Direct-seed snap beans, sweet corn, squash, and cucumbers.
- ☐ Late this month, start seeds of broccoli, kale, and cauliflower indoors.
- ☐ Toward month's end, sow peas, beets, and carrots in the garden.
- ☐ Direct-seed cold-hardy herbs such as parsley, chives, and sage.
- ☐ Have *Bacillus thuringiensis* (BT) ready to use if leaf-eating caterpillars show up on brassicas.

Zone 10

- ☐ Plant okra early this month.
- ☐ Set out transplants of tomatoes, peppers, and onions.
- ☐ Direct-seed cucumbers, melons, and squash, as well as herbs.
- ☐ Replenish mulches and soil amendments, such as compost.
- ☐ Prune poinsettias for holiday bloom.

Pick No Produce Before Its Prime!

Nearly every vegetable and fruit provides telltale cues that indicate ripeness. Here's what to look for:

Apples

Pick apples when fully colored and the skins begin to lose their shine. At that stage, the fruits' flavor will be perfectly balanced.

Beans

Begin picking beans when the seeds just start to show through the first pods. After that, harvest at least every other day so that the beans don't become tough, stringy, and overmature. And, if possible, harvest first thing in the morning.

Bell Peppers

For maximum nutrition and sweet flavor, allow peppers to mature to their fully ripe color (red, gold, purple, etc.). If you prefer the flavor of green peppers, wait for the peppers to reach full size and change color from light to dark green; at this stage, most peppers lose their bitterness.

Brambles and Blueberries

Ripe raspberries and blackberries will pull easily off the canes. Blackberries come off with the core intact, while raspberries pull away from the core. If you have to tug to remove the berry, leave it alone—it's not ripe yet.

If you have lots of blueberry bushes, you can place a tarp beneath the plants and gently shake them at harvest time, as large-scale commercial berry growers do. Ripe berries will fall onto the tarp. For smaller plantings, take the time to inspect the plants, and pick only those berries that are totally blue. Berries that have even tiny green spots on them will be less sweet than fully ripened blueberries.

Broccoli and Cauliflower

For finest flavor and texture, harvest broccoli and cauliflower when the heads are still tight and hard. If you're not going to use them immediately, stick the stem ends in a pitcher of water and store in the refrigerator for up to 1 week.

Carrots

A carrot's color is a good indicator of ripeness. Scratch away a bit of soil from the top of one side of the carrot. Harvest roots that are dark orange, not pale.

Cherries

You can't judge a ripe cherry by its color! Instead, gently press the fruits with your thumb. If they're hard, wait a few more days and test again. When cherries feel soft and plump, they'll be sweet and juicy. And by the way, don't worry about whether or not the cherry stems are attached to the fruits when you pick. The fruits' quality doesn't seem to be affected either way.

Cucumbers

The best measure of a ripe cuke is its size. Pick slicing cucumbers when they are about 6 inches long and firm. Pick picklers when they're about 4 or 5 inches long. If you see any yellow at all, you've missed the boat.

Grapes

Harvest grapes when they are fully colored. Some varieties (such as 'Niagara' and 'Concord') also become intensely aromatic when they are ripe—you'll notice their fragrance from almost 10 feet away.

Harvest low-acid varieties, such as 'Van Buren', as soon as they ripen. If left too long on the vine, these grapes lose flavor and texture. Highly acid varieties, such as 'Baco Noir', can hang on the vine a bit longer until the grapes sweeten and lose some of their acidity.

Melons

Harvest melons at their "full slip" stage, when the fruits separate (or slip) easily from the vine without tugging. Also, the stem that connects the fruit to the vine will be completely dry.

On cantaloupes, inspect the long stem with a small leaf on it, at the point where the melon joins the vine. When the melon is ripe, that leaf will be paler than the other leaves on the vine. On a ripe watermelon, the tendril will be withered and dark.

Peaches

Peaches ripen from the bottom up and from the smooth side to the creased side. To test for ripeness, gently press on the upper shoulders near the crease. If it gives a little, the peach is ripe and ready to pick.

Pick sweet corn when silks dry and turn dark brown to within ½ inch of the tips of ears.

Peas

Pick peas before the top of the pod dries out completely. The pod should look plump and the stem end will have a few small white flakes, signaling that the pod is beginning to dry. If you're still in doubt, open one or two of the pods to check the peas' size inside. They're perfect for picking when plump and nicely rounded.

Potatoes

For tender "new potatoes," dig up your spuds about 3 weeks after the flowers first appear. For larger potatoes suitable for storage, harvest after the tops of the plants die back.

Strawberries

Ripe strawberries turn dark red, usually just a day after they become bright red. Waiting that extra day to harvest really enhances the berries' sweetness. Check the underside of each berry before picking to be sure that it has ripened all the way through. In fact, if you have a small patch, you can gently turn the berries as they grow to help them ripen more uniformly. To keep strawberry plants producing, harvest every other day when the berries are ripening.

Sweet Corn

Your corn is ready to harvest when the silks have dried to within ½ inch of the tips of the ears. Still not sure? Peel back the husk at the tip of the ear, then use your thumbnail to puncture a kernel near the top. On ripe ears, a milky liquid will squirt out. (Super-sweet varieties, an exception, exude clear liquid even when ripe.)

Tomatoes

For the most-flavorful tomatoes, harvest the fruits just as they begin to change color from orange to red. Set the tomatoes on a counter (not a sunny windowsill) with normal indoor light and at room temperature for 4 or 5 days, until they turn fully red. Never store tomatoes in the refrigerator; temperatures below 50°F destroy tomato flavor and texture.

Winter Squash

Wait until the stem begins to turn brown and the skin is fully colored. On buttercup types, the skin will turn solid dark green. Cure the fruits in a cool, dark place for about 1 month before you eat them.

Preserve Your Harvest the Cool Way

Putting up your bounty by canning in a boiling water bath can seem tedious and time consuming, especially when plenty of work still remains in the garden. So why not take the cool route by *freezing* your favorite vegetables and fruits? Freezing is quick and easy. And even if you are limited to the small freezer compartment of your refrigerator, you still can freeze a few packages to help you get through winter's dark days.

With the exception of leafy vegetables (such as lettuce, spinach, and cabbage), most vegetables and fruits freeze well.

Blanching Matters

Harvest your crops when they are young and tender, then freeze only the best-looking, freshest fruits and veggies. Overripe fruits tend to deteriorate when frozen.

Before you freeze your vegetables, blanch them by boiling or steaming for a short time, then cooling them in ice water. Blanching inactivates enzymes that would otherwise cause the loss of texture, color, flavor, and nutrients. Although you can safely eat frozen vegetables that haven't been blanched, the quality will be poor.

Studies at the University of Illinois found that vegetables blanched before freezing retained more of their A, B, and C vitamins than those put directly into cold storage. For example, after a month in the freezer, blanched peas retained 11 percent more of their vitamin C than unblanched ones. Three months after freezing, that difference climbed to

Perfect Pea Preservation

Freezing peas with edible pods is a snap. Stringless varieties, such as 'Sugar Daddy', work best because they don't fall apart during processing the way stringed varieties do.

For freezing, harvest snap peas when the pods are only half full. As soon as possible after picking, rinse the pods in cold water, then snap or cut off the ends where they connect to the vine. Blanch small pods for 2 minutes and large pods for 3 minutes in boiling water, or steam small pods for 4 minutes, large pods for 5. Plunge the blanched peas into ice water to chill them quickly. Drain, then pack the peas into airtight containers and freeze them.

Without blanching, frozen vegetables lose flavor, texture, and nutrients. For best flavor, freeze snap peas when the pods are only partially filled.

(see "Blanching Basics" on this page)

Blanching Basics

Blanching helps preserve produce color and texture by stopping the enzymes that cause breakdown during storage. To blanch, just drop a pound or so of veggies into boiling water, or steam them. After the required time, remove the vegetables and plunge them immediately into ice water until they have cooled. Here are the blanching times required for common veggies:

Beans, green shell:
   1 minute

Beans, snap and lima:
   2 or 3 minutes

Broccoli (florets):
   1 to 3 minutes

Brussels spouts:
   3 to 5 minutes

Carrots: 2 minutes

Cauliflower (florets):
   3 minutes

Corn: 3 or 4 minutes

Okra: 3 or 4 minutes

Peas, edible podded:
   2 or 3 minutes

Peas, shelled:
   1 or 2 minutes

Summer squash (slices):
   3 minutes

36 percent; and by the ninth month, the blanched peas had 89 percent more vitamin C!

And since blanching is quick and easy, there's no reason not to do it. Set up your kitchen in an assembly line fashion and you'll be finished in no time. Boil water in large pots or set up steamers on the stove, put another large pot filled with ice and cold water in the sink, and line up your freezing containers on the counter.

Just before blanching, clean the vegetables thoroughly and remove inedible stems, strings, husks, and pods. For water blanching, boil 1 gallon of water for each pound of vegetables. To steam, use a steamer, colander, or wire-mesh basket over 1 inch of water boiling in an 8-quart pot. Don't use the microwave oven for blanching—it may not stop the breakdown by the enzymes. Each vegetable has its own time requirement for blanching (see "Blanching Basics" on this page); adhere to the recommended times as closely as possible for the best results. If you're blanching in boiling water, start timing when the water returns to a boil after you've added the vegetables.

To stop the blanching process, plunge your veggies into an ice-water bath. Never put warm foods in the freezer—it causes large ice crystals to form. As a rule of thumb, cool your vegetables for the same amount of time as you blanched them.

Packing Pointers

After cooling your veggies, drain them, then pack them loosely into containers designed for the freezer. Cottage cheese cartons, bread bags, and other reused packages are poor choices because they are not made specifically for freezing. If they're not airtight, your produce can get freezer burn. Glass jars also are a "no-no" unless they're the type designed for freezing. Mayonnaise and other reused glass jars easily crack when exposed to sudden temperature changes. Resealable plastic freezer bags may work the best because they take up less room in the freezer than plastic freezer cartons.

When filling containers, remember that foods expand as they freeze. Leave ½ inch of space at the top of plastic cartons, and don't overfill plastic bags. Before you seal the containers, press or squeeze out air pockets.

When you're ready to use your frozen harvest, take the vegetables from the freezer and drop them directly into a steamer. If you thaw frozen vegetables before cooking them, condensation will seep in and dilute their flavor.

One last piece of advice: Don't overcook your frozen veggies! Blanched and frozen vegetables will be ready to eat in about half the time that fresh ones would be.

Technique

Freezing Sweet Corn: Step by Step

Corn on the cob is the vegetable that most perfectly evokes the sultry sensation of a late-summer supper. To capture that fresh summer flavor without taking up all of your freezer space, cut the kernels off the cob before freezing. With the help of a friend or two, you can freeze several hundred ears in just a few hours. Here's what to do:

1. Husk freshly harvested ears just before you're ready to freeze them. Super-sweet varieties hold their flavor and texture especially well when frozen.

2. Working with about 6 to 8 ears at a time, drop the corn into a 12-quart pot of boiling water. Blanch each potful of ears for 3 or 4 minutes.

3. Use tongs to lift the ears out of the boiling water, then drop them immediately into another container filled with ice water. Let the ears cool for another 3 or 4 minutes.

4. To remove kernels from a cob, hold the ear upright in a bowl and run a paring knife down the cob. The kernels will fall off in strips. Work your way around the ear until you've removed all of the kernels.

5. Use a strainer to scoop the cut corn into plastic freezer bags. Seal the bags, then pop them into the freezer.

6. To cook your corn, remove the frozen kernels from the freezer bag and put them directly into a pot on the stove. Add a small amount of water or butter to keep the corn from sticking. Cook the corn just long enough to thaw and heat it—it needs no extra cooking.

Use a sharp paring knife and work in long, downward strokes to remove kernels from a cob.

Freeze Tomatoes and Peppers in a Flash

Unlike most other vegetables, tomatoes and peppers *don't* need to be blanched before they're frozen.

To freeze unpeeled whole tomatoes, simply wash and core the fruits, then pack them in freezer bags. When you're ready to use them, put the frozen bag under hot water briefly until you can remove the tomatoes. Stick the tomatoes under hot water for a few more seconds to loosen the skin, which will easily peel off.

To freeze peppers, just chop or slice them, freeze the pieces on a cookie sheet until they're solid, then transfer the pieces to a plastic freezer bag. When you're cooking, just scoop out the amount of peppers you need.

Storage Requirements of Vegetables and Fruits

COLD AND VERY MOIST (32°–40°F; 90–95% RELATIVE HUMIDITY)	Beets Broccoli (short term) Brussels sprouts (short term)	Carrots Celery Chinese cabbage Collards	Leeks Parsnips Rutabagas Turnips
COOL AND MOIST (40°–50°F; 85–90% RELATIVE HUMIDITY)	Cucumbers Eggplants (50°–60°F) Muskmelons (cantaloupes)	Peppers, sweet (45°–55°F) Tomatoes, ripe Watermelons	
COLD AND MOIST (32°–40°F; 80–90% RELATIVE HUMIDITY)	Apples Cabbage Cauliflower (short term)	Grapefruit Grapes (40°F) Oranges Pears	Potatoes Quinces
COOL AND DRY (35°–40°F; 60–70% RELATIVE HUMIDITY)	Garlic Onions Soybeans, green, in the pod (short term)		
MODERATELY WARM AND DRY (50°–60°F; 60–70% RELATIVE HUMIDITY)	Peppers, hot (dried) Pumpkins Squash, winter	Sweet potatoes Tomatoes, green (tolerate up to 70°F)	

Signs of Spoilage

If you suspect that your canned fruit or vegetables have spoiled, don't even bother opening your preserved pickings. Doing so could release a bacterium so toxic that a taste may be fatal. Place jars containing spoiled produce in boiling water for 30 minutes, then wrap the jars in several layers of plastic bags, sealing them well, and dispose of them where they can't be found by animals or children; do not empty the contents down the sink, garbage disposal, or toilet because they could contaminate the water supply. If the contents happen to escape, be sure to sanitize your hands, counter, and utensils with hot, soapy water.

Check every jar before you use it, looking for the following signs of serious contamination:

◆ A jar that is soiled or moldy on the outside indicates that food has seeped out during storage, which means that air, bacteria, yeasts, and molds could have gotten in. Jars right out of the canner might be a bit soiled from some of the liquid that was drawn out with the air. This is okay as long as half the contents of the jars aren't floating outside in the canning water. If you've wiped the jars as you should have done, the jars would have gone into storage clean, and any food on the outside of the jars now is not okay.

◆ A significant change of color, most notably a much darker color, can mean spoilage. Some brown, black, or gray discoloring may be due to minerals in the water or in the cooking utensils; although it may detract from the looks of the food, there is no harm done otherwise.

◆ A change in texture, especially if the food feels slimy, is a sure sign that the food isn't fit to eat.

◆ Mold in the food or inside the lid—sometimes nothing more than little flecks—is not a good sign.

◆ Small bubbles in the liquid or a release of gas, however slight, when you open the jar means foul play. Sometimes you get a strong message: Liquid actually spurts out when you release the seal. Other times the gas is more subtle.

◆ A swollen lid indicates the presence of spoilage yeasts and bacteria that have formed gases inside the container.

Pantry Records

ITEM	HOW MUCH	STOCKING-UP METHOD	NOTES

GARDENER TO GARDENER

Effortless Walnut Shelling

I discovered the secret to shelling walnuts so that you get perfectly formed halves—not tiny fragments like the ones you get from a conventional nutcracker. Here's how to remove the brittle shells with little effort:

1. Place a solid piece of iron on a table. (I use a wood-splitting wedge.)

2. Hold the walnut upright on the iron with the nut's pointed end up.

3. Strike this point solidly with a hammer, but not so heavily as to shatter the nut.

4. Once the point of the nut has been indented, the shell will have vertical cracks and sections of it will come away like the peelings of an orange.

If the nut shatters, swing the hammer more gently the next time. With a little practice, you'll know the proper strength to use. This method has proved faster, easier, and more productive than any of the three patented nutcrackers I've used over my long walnut-cracking career.

Arnold Rothmeier
Gridley, California

Wash-and-Wear Carrots

After harvesting a bumper crop of baby carrots, I began looking for an easy way to clean them. I didn't relish the thought of standing over the sink all day, so I took a more mechanized approach: After cutting off the tops of the carrots, I tossed the roots in my washing machine, filled it with cold water, and turned it on. Half an hour later (spin cycle and all), I had a tubful of beautifully clean baby carrots. I packed them in plastic vegetable bags and put them in the fridge for my family to snack on. It worked great!

Lolly Osguthorpe
Orem, Utah

Have Some Kaleoli

Kale has become one of our family favorites. Here in eastern North Carolina, we pick its nutritious green leaves straight through winter and into April. One year we discovered a bonus: When the weather warms up, kale produces flower buds similar to broccoli. These buds make an excellent, tangy addition to a vegetable stir-fry. When the unpicked buds open on the plants, we cut the flowers and add them to spring bouquets. Then, we save the seed to start again in late summer for a repeat performance the following year. What more could you ask from a vegetable that requires such little care?

Marion D. Friedlander
New Bern, North Carolina

Celebrate the Season

Each year we put out a Thanksgiving spread that includes the absolute best of our harvest that year. In June, for instance, we freeze our sweetest strawberries in bags marked "Do not open until Thanksgiving." Similar labels adorn special jars of homegrown/homemade preserves on our pantry shelves and even mark the finest winter squash and pumpkins stored in the cellar. Knowing that we helped nature bring forth the bounty on our holiday table makes Thanksgiving an extra-special day for our family.

Jim Busch
White Oak, Pennsylvania

GARDENER TO GARDENER

Harvest Centerpieces

Many people decorate their homes with fall flower arrangements, but few realize that fruits and vegetables from the garden or market make lovely fall designs and centerpieces, too. Try working colorful Swiss chard, peppers, corn, and cherry tomatoes into your arrangements. How about a bunch of trimmed leeks, bundled together with a decorative raffia bow? For added interest you can shred the stalks with scissors and poke fresh flowers into the foliage.

For a really unusual and striking centerpiece, turn a fresh head of cabbage into a vase. Just hollow out the center and insert a piece of floral foam into the space. Moisten the foam and add flowers, grasses, or even unusual fruits to complete your arrangement.

Joe Helmick
Bexley, Ohio

Easy Sunflower Seed Hulling

When you need to shell a large amount of sunflower seeds, such as for baking, try using a chipper/shredder! It's a lot faster and easier than hand shelling.

I discovered this method by accident while shredding garden refuse in my electric shredder. First, I sent through the sunflower stalks. Then I cut the flowerheads (with a few seeds still left on them) into large pieces and began shredding them. Soon I noticed whole, cleaned seeds in my bucket. After this discovery, I cleaned the shredder thoroughly and poured in the rest of my unhulled seeds. To clean off the remaining debris, I winnowed the seeds into a bowl in front of a box fan. In almost no time at all, I had 2 cups of cleaned seeds ready for baking.

Duane Howard
Grand Junction, Colorado

Editor's note: Be sure to clean your shredder thoroughly before you do this.

Fruitcake Alternative

Homegrown fruits can be dried and mixed to give as special gifts for family and friends. My fruit mix consists of plums, apricots, peaches, nectarines, apples, pears, figs, and oranges. I dry the fruits in a food dehydrator as each type ripens in my garden. When the fruits have dried, I put them into wide-mouth 1-quart canning jars and store the jars in my refrigerator. (The maker of my dehydrator says the fruits will last up to 24 months at 34°F, and up to 8 years at 0°F, but I haven't been able to verify these claims. My fruits are always eaten long before then!) To make the gifts special, I attach a ribbon and poem to each filled jar.

Robert Lestak
Moorpark, California

Chapter 10

October

Stretching the Season: Predicting Frost, Protecting Plants

The frost performs its secret ministry, unhelped by any wind. . .

—*Samuel Taylor Coleridge* ("Frost at Midnight")

With days growing shorter, leaves changing color, and nights becoming cooler, there can be little doubt that Nature is preparing to kick back for a while. For most of us, that means the end of the growing season and, along with it, our harvest of food and flowers.

As organic gardeners, our job is to work with Nature—at least most of the time. When it comes to eking out a few extra weeks of growing and harvest, many gardeners pull out all the stops in an effort to forestall the inevitable. Knowing exactly when to expect that first frost in our gardens gives us an edge, allowing us to protect our plants from killing cold—even when the weather forecast for the nearest city assures frost-free conditions. Then, when frost looks likely, we round up our blankets and ready our coldframes, hoop houses, and greenhouses. With the right equipment, we can *persuade* Nature to give us just a little more time.

Lucky gardeners in the Deep South's Zones 9 and 10, however, don't need to go to these lengths to keep their harvest going. October's cooler temperatures are just right for growing salad greens, brassicas, and root crops—no special devices or techniques required!

Gardener's To-Do List—October

**If you don't know what USDA hardiness zone you live in,
check the map on page 230 to find out.**

Zone 3

- [ ] Clean up the garden and compost the residue.
- [ ] Finish harvesting cauliflower, cabbage, broccoli, beets, and turnips.
- [ ] Cover hardy crops, such as spinach, parsley, and kale, to keep them growing another month.
- [ ] Shred leaves, then use them for mulch.
- [ ] After the first hard freeze, cover perennials with a winter mulch of leaves, pine needles, or straw.
- [ ] Wrap the bases of fruit trees with chicken wire to protect against winter rodents.

Zone 4

- [ ] Time to harvest brussels sprouts, kale, parsley, and other hardy crops, or to prolong the harvest by covering them on cold winter nights.
- [ ] Store carrots in the garden all winter by removing the tops, then covering the bed with a foot of mulch.
- [ ] Brush white paint on young tree trunks to protect them from sunscald.
- [ ] Rake up fallen leaves, then mow or shred them. Wait until the ground freezes to use them as mulch over peonies, lilies, and other perennials.
- [ ] Scout your local garden center for bargain bulbs, then get them into the ground before month's end.

Zone 5

- [ ] Plant annual rye or winter wheat in vacant beds to prevent weeds and provide "green manure" for next season.
- [ ] Mulch carrots heavily to keep them from freezing.
- [ ] Cover salad greens with plastic on frosty nights.
- [ ] Harvest kale and brussels sprouts.
- [ ] Prune roses, rake up all their old mulch, then replace it by piling fresh straw or chopped leaves up to the lowest buds.
- [ ] Rake the orchard floor clean to interrupt the life cycles of pests and diseases.

Zone 6

- [ ] Gather up and compost all your withered garden plants.
- [ ] Harvest broccoli, cauliflower, cabbage, and kohlrabi as soon as they're ready, or be prepared to cover them on cold nights.
- [ ] Dig up sweet potatoes and harvest pumpkins and winter squash before the first hard freeze.
- [ ] Weed blueberries and raspberries, then mulch them with shredded leaves.
- [ ] Dig up and store tender bulbs and tubers, such as cannas, begonias, and caladiums, before the first frost.

- ☐ Cut back chrysanthemums and asters after the flowers fade, but wait until spring to move them.
- ☐ Plant spring-blooming bulbs.
- ☐ Set out new strawberries.

Zone 7

- ☐ Thin any spinach that you won't overwinter, and eat the thinnings.
- ☐ Cover broccoli and cauliflower on frosty nights.
- ☐ Plant garlic and multiplier onions.
- ☐ Set out evergreen shrubs.
- ☐ Sow seeds of poppies, larkspur, blue flax, hollyhock, bachelor's buttons, and sweet rocket.
- ☐ Plant spring-flowering bulbs.
- ☐ Set out new strawberries.
- ☐ Stockpile leaves and pine needles.

Zone 8

- ☐ Sow more spinach and parsley, and thin leafy greens.
- ☐ Dig up sweet potatoes and peanuts, and cure them before storing.
- ☐ Keep the roots of tomato plants constantly moist; ripening fruits are less likely to crack after heavy rains.

- ☐ Plant bunching onions, regular onions, and leeks, and prepare a fertile spot for garlic.
- ☐ Replace summer flowers with hardy annuals.
- ☐ Lift and store caladiums.
- ☐ Go wild by direct-seeding wildflowers, such as gilia, black-eyed Susan, poppies, candytuft, coneflowers, and coreopsis.

Zone 9

- ☐ Direct-seed Chinese cabbage, spinach, lettuce, and radishes.
- ☐ Set out seedlings of the cabbage-family crops you started last month, and check them daily for pests.
- ☐ Harvest sweet potatoes, cure them, then store them.
- ☐ Color-up fall flowerbeds with alyssum, calendula, dianthus, snapdragons, and ornamental kale.

Zone 10

- ☐ Thin and water fall greens and root crops.
- ☐ Fill empty spaces in the garden with Chinese cabbage, beets, cauliflower, collards, broccoli, kale, turnips, and spinach.
- ☐ Set out new strawberry plants.
- ☐ Lightly fertilize citrus trees with compost or other organic fertilizer, and water them if rainfall has been low.

When Frost Threatens

After reading all the indicators, you determine that frost may hit your garden tonight. What can you do?

- **Cover plants to retain warmth and moisture**. Use old sheets or blankets, pine branches, straw, inverted pots, or water-filled cloches.

- **Protect lettuce, arugula, chard, beets, and mustard from the wind** and they will survive near-freezing temperatures.

- **Mulch carrots and other root crops before a frost** to keep the ground from freezing hard, then harvest as needed.

- **Dig up tomato plants and hang the vines indoors**. If you keep them warmer than 60°F, the remaining fruits will ripen.

- **Pot up peppers and bring them indoors**. Remove the immature peppers and cut back the stems.

Predicting the First Frost

It will happen any day now. Clear sky. Bright sunshine. Low humidity. And your local forecaster will say, "Possible frost tonight." But will the frost hit *your* garden? It's something gardeners fret about, especially when the threat comes just a week or two after planting.

All weather is local—so local that your neighbor's garden just 100 yards away may suffer a damaging frost while yours doesn't. But if you know the lay of your land and a few easily observed atmospheric clues, you'll be able to predict exactly when the first frost will hit your garden. With this information, you can plan to protect some plants and get others ready for the compost pile.

Clouds

If the sky is clear, the air is dry, and the temperature is falling, chances are frost will settle on your garden. But if the sky is cloudy, frost is less likely because low, thick clouds will act like a blanket. They prevent the heat that the soil radiates at night from escaping into the atmosphere, keeping frost from settling on your plants.

Breezes

During the day, the soil absorbs heat from the sun. At night, it radiates that heat back into the atmosphere, and cold air gradually settles around your plants. But a slight breeze mixes the somewhat warmer air from above with the cold air near the soil, raising the air temperature a few degrees. So frost is less likely on a night with a gentle breeze (but not a strong wind).

Frost Pocket

Cold air is dense. The molecules are packed tightly together, and the air is so heavy that it flows downhill and, like water, pools in low places. A valley may be as much as 18°F cooler than the adjoining slope. Likewise, the temperature in your garden varies from a sunny slope near a wall to the lowest point.

Southern Slope

A gentle slope facing south receives solar radiation—heat and light—longer than other sites. And the radiation is more intense. That's why a southern slope is the best location for a garden. Also, cold air drains down slopes (as described above), so gardens on top of slopes will get frost later than those at the bottom or on level spots.

Dew Point

As the evening temperature falls, the air can hold less and less moisture until it reaches the *dew point*, at which the moisture condenses as dew. Heat is released in the process, helping to keep the air temperature at or only slightly below the dew point. So the more moisture in the air at sunset, the less the likelihood that frost will occur during the night. (This is why commercial growers turn on sprinklers when frost is predicted. The added moisture in the air raises the dew point.)

Some forecasters give the dew point in their reports. If yours doesn't, call the local National Weather Service office, or visit the NWS Web site at www.nws.noaa.gov.

Trees

Trees act like a blanket that prevents ground heat from escaping into the atmosphere. They also exude moisture, raising the dew point and reducing the chance of frost. A large vegetable garden surrounded on three sides by mature oaks often can survive the first two or three frosts untouched.

Double Covers Keep Greens Growing

When frost threatens, an extra blanket may be all you need to keep greens growing well into fall and even winter. Simply use two layers of fabric row cover to protect lettuce, spinach, mustard, and other greens. Anchor the edges in place with soil, rocks, or whatever you have on hand. This double cover-up also works great for warming crops in early spring.

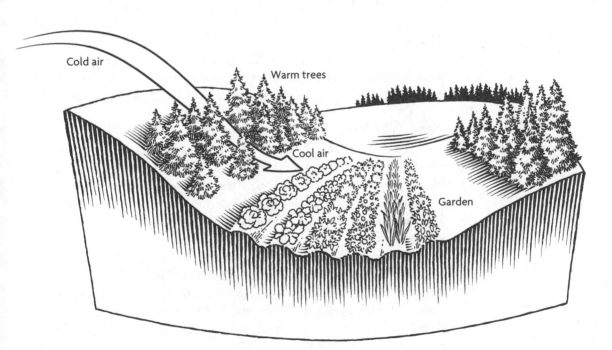

Heavy cold air flows downhill and settles in low places, sometimes called "frost pockets." Trees on the slope will warm the draining air, reducing the likelihood of frost in the hollow.

Walls

Cold air rushing down a slope collects not only in hollows but also behind stone walls, fences, and rows of dense vegetation such as hedges. Frost occurs sooner at the base of these barriers. To avoid frost damage beside such an obstruction, provide an opening in it through which the cold air can drain.

On the other hand, a south-facing stone wall is a "heat sink." During the day, the sun warms the stones, which release the heat at night, making the plants on the south side of the wall less prone to frost.

Windbreaks

A gentle breeze tempers ground-hugging cold air by mixing in warm air from above, protecting your plants from frost. But heavier winds dry out plants, reducing the humidity around them and making them more susceptible to frost. A windbreak can reduce the wind's effects by 50 percent and intensify heat and light by reflection.

Four or five rows of deciduous shrubs, or two or three rows of evergreens, make an effective windbreak. For maximum effect, the height of the windbreak's plants should increase from windward to leeward, so that the tallest plants are near the ones you're trying to protect.

Ponds

Bodies of water near your garden—an ocean, a lake, or even a farm pond—will moderate temperatures. It takes five times as much heat to warm a body of water 1°F as it does to warm an equal quantity of soil.

Your Garden

Dark surfaces absorb more heat than light ones, and dark, fertile soil stays warm longer. Deep soil retains moisture longer. And clay soil retains more heat than sandy soil. These may seem like minor influences, but even a tenth of a degree of warmth can make a crucial difference between enjoying another few weeks of growth and raking up blackened stems.

Low, compacted plants expose less of their leaf area to wind, so bush crops are less vulnerable to frost than tall vining plants. Dark-colored plants absorb more heat than light ones and are more likely to survive frost. Hairy leaves also retain warmth. Plants growing close together share heat with each other. A dense planting also interrupts the upward movement of heat from the soil, holding it in the garden.

Mulch prevents soil heat from rising to warm the air around your plants, so a garden with a thick layer of mulch will succumb to frost before a garden with little or no mulch.

Instant Coldframe

Keep an eye out for treasures in your neighbors' trash! A salvaged skylight can serve as an excellent ready-made coldframe for lettuce, spinach, Oriental greens, perennial herbs, and other cold-hardy plants. The ideal model is a one-piece dome made from a double layer of Plexiglas—perfect for letting in light and keeping out cold. Simply build a rectangular wooden frame to support the covering.

Technique

Build a Raised-Bed Hoop House

A hoop house is an inexpensive, portable structure for extending the growing season. For extra frost protection, you can cover it with an agricultural foam "blanket" on cold nights. Here's how to build a simple 4-foot-high hoop house:

1. Build a raised-bed frame (about 4 feet wide × 8 feet long) using untreated landscape timbers. Drill four equidistant 1-inch-diameter holes into the tops of both long sides, drilling all the way through the wood. These holes will anchor the hoops.

2. Create a template for the hoops by driving wooden stakes into the ground, forming a 4½-foot-high arch. The two bottom stakes should be about 4 feet apart, with the curved section beginning about 21 inches from the bottom of each side. Arch and form the hoops using the ¾-inch-diameter metal conduit sold to protect electrical wires. Gently bend the conduit around the stakes.

3. To raise the roof, drill four equidistant 1-inch-diameter holes in the ridgepole to match the four pairs of holes in the frame. Run each hoop through a hole in the pole, securing it by running a sheet metal screw down through the pole and into the hoop. Insert the hoops into the frame.

4. Trim the plastic squares to be about 1 inch bigger all around than the area of the end hoops. At each end, put one piece on the inside of the hoop and one on the outside, then staple the edges together all around the hoop.

Center the large piece of plastic over the ridgepole, then staple it to the pole. (To keep the plastic from tearing, cut thin cardboard into long strips and place these over the plastic before stapling.)

On each side, anchor the plastic by stapling the bottom to the two remaining pieces of lumber. When ventilation is needed on warm days, simply roll up the sides.

Materials

Frame: Untreated 6" × 6" landscape timbers (two 8' pieces and two 4' pieces)

Hoops: ¾"-diameter flexible conduit (four 10' pieces)

Ridgepole: One 8' 2 × 2 stud

Cover: One 10 × 25' roll of 6-mil greenhouse-grade plastic (cut four 52"-square pieces and one 8½ × 12½' piece)

Side anchors: Two 8' 2 × 2 studs

Also: Small wooden stakes, staples, sheet metal screws, cardboard

The roll-up sides on this simple hoop house allow ventilation on warm days, as well as easy access to plants.

Greenhouse Seed Starting

To start seeds in a cool, unheated greenhouse, use a germination mat—a rubber heating pad that provides consistent (and in some models, adjustable) bottom heat. Stack your seed-starting flats so that heat-loving seeds, such as hot peppers, are on the bottom, closest to the heat. Put cool-loving germinators, such as lettuce, above them, where temperatures will be 5° to 10°F cooler.

Or, simply germinate the seeds in your house, where it's already about 70° F. After the seeds sprout, transfer the flats to the greenhouse, where the light is brighter.

Green Thumb? Get a Greenhouse

Want to extend your growing season to include all four seasons? Buy or build a greenhouse. Many sizes and styles are available, ranging from simple, unheated structures that you fit on a porch or patio to freestanding, climate-controlled conservatories. No matter what type of greenhouse you use, paying attention to these factors will maximize your chance of success:

1. Get the light right.

The quantity and quality of the light that enters a greenhouse from the outside can vary tremendously depending on the type of see-through material—known as *glazing*—that is used for the ceiling and walls. Corrugated polyvinyl chloride (PVC), for example, lets in 84 percent of the available light, while Plexiglas lets in 93 percent. That may not seem like a big difference, but it is to your plants!

As a rule of thumb, every 1 percent increase in light transmission of your glazing material increases plant growth 1 percent during winter. So, that 9-percent difference between PVC and Plexiglas could well prove to be the difference between success and failure—especially if your greenhouse will be located in an area with a lot of winter cloud cover. There, greenhouse owners need to make the most of every bit of sun they get.

Cleanliness counts, too. If your greenhouse covering is dirty or dusty, wash it. Then establish a regular schedule of window washing.

If your greenhouse is covered with a flexible plastic film that has yellowed, replace it with a greenhouse-grade plastic designed to take the sun (available from greenhouse supply companies). The newer greenhouse plastics contain ultraviolet inhibitors that can extend the life of the plastic 4 years; infrared inhibitors, which reduce heat loss at night; and wetting agents so that condensation runs down the sides of the greenhouse instead of dripping onto your plants.

How you arrange plants inside the greenhouse can make a difference too. The area of brightest light is next to the glazing. At the end walls, where the framing interferes with light transmission, the lower light levels are better for shade-loving plants like impatiens.

2. Keep it hot, hot, hot . . .

Greenhouses are like growing zones. They contain microclimates of cold and hot spots. If you learn where these areas are, you can satisfy the needs of both heat lovers (such as tomatoes, peppers, eggplants, and cucumbers) and cool-clime plants (such as lettuce)—all in the same greenhouse.

Place several thermometers throughout your greenhouse to learn the best place for growing specific plants. Shade-tolerant azaleas, for instance, may do best under the bench, where the temperature is a bit cooler and the light is indirect. Tropical orchids and bougainvillea are likely to thrive near the ceiling, where it's hotter and brighter.

Just be aware that the warmest spot by day—next to the glazing— becomes the coldest spot at night. You can stabilize indoor temperatures by incorporating heat-storing masses, such as brick floors and water barrels, in your greenhouse's design.

You may also want to add some supplemental heat to your greenhouse. If the greenhouse is attached to your home, you can tie it into your home heating system. If you choose to use a natural gas or propane heater, be sure to spend the extra money to vent the heater outside. Both of these gases produce sulfur dioxide and ethylene gas, which can harm plants. The sulfur dioxide, for instance, combines with the water that condenses on the inside of glazing to form sulfuric acid (acid rain!). Be careful, though, that you don't melt the glazing with the heater's surface.

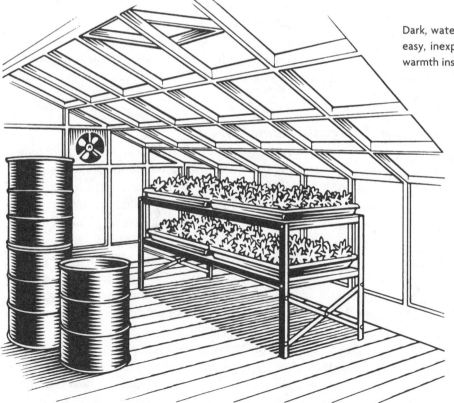

Dark, water-filled barrels are an easy, inexpensive way to retain warmth inside a greenhouse.

For More Information . . .

Greenhouses can be attached to your home or can stand alone. Your choice will depend on many factors, including your house design and gardening needs.

For more information on buying and using a greenhouse, contact the Hobby Greenhouse Association, 8 Glen Terrace, Bedford, MA 01730-2048; www.hobbygreenhouse.org. Members are knowledgeable and ready to help!

3. Keep the air moving.

Warming a greenhouse is important, but keeping it cool can be crucial! In many parts of the country, too much heat is the biggest problem greenhouse growers must prepare for. As the sun beats down on your greenhouse during the unpredictable weather of spring, the air trapped inside can quickly heat up to 110°F or more. Ceiling vents that let hot air escape from the top of the greenhouse are the best protection, particularly in hot or sunny climates. Second best are vents and fans located on the end walls to bring in cool air from outside.

If you aren't around during the day to control the ventilation yourself, spend a bit more to get automatic vents. Most are gas-filled cylinders that push the vent open as rising heat expands the gas inside. You also can get thermostatically controlled exhaust fans that run when the mercury reaches a preset temperature.

Air circulation is important even when you aren't venting excess heat. Having a fan running continuously inside a sealed greenhouse, for instance, is nearly a necessity. It can lessen the chance of plant diseases taking hold and help prevent depletion of the carbon dioxide (CO_2) that plants need for photosynthesis.

Individual plants tend to totally deplete the CO_2 in a tiny envelope of air that surrounds each leaf. But the CO_2 in these tiny pockets can be replenished somewhat by simply moving the air around with a fan. To completely replenish the CO_2, you'll need to open a vent to bring in fresh air.

4. Gauge the humidity.

Buy a humidity gauge (you can find them in hardware or home supply stores) so that you can maintain a relative humidity level of 40 to 50 percent inside your greenhouse. Most times, this means adding a little moisture to the dry winter air. In that case, either run a humidifier or try venting your electric clothes dryer into the greenhouse.

But you could need to *lower* the humidity, instead. If you overwater (especially in winter), the excess will go right into the air, raising the humidity level to as high as 95 percent. And high humidity coupled with poor air circulation can set off an explosion of root rot or powdery mildew on your plants.

To prevent such problems, water only when the soil in your greenhouse containers is dry—and watch that humidity gauge. You can also run an exhaust fan or just open a window on a warm winter day to decrease the humidity.

Top **10**
Greenhouse Vegetables

When adapting garden vegetables to a greenhouse, keep each plant's natural growing season in mind. Listed below are a few of the more hardy greenhouse plants and the best times to plant them. Most of the following plants will perform well with artificial lights and a good heater, though some may be more prone to insect and disease problems.

1. **Beet greens:** Plant mid-February to May, then again in August to mid-October.

2. **Eggplant:** Plant mid-February to March.

3. **Herbs:** Plant mid-February to mid-October.

4. **Lettuce:** Plant mid-February to May, then again in August to mid-October.

5. **Mustard:** Plant mid-February to May; then again in August to mid-October.

6. **Oriental greens:** Plant mid-February to May, then again in August to mid-October.

7. **Peppers:** Plant mid-February to March.

8. **Spinach:** Plant mid-February to May, then again in August to mid-October.

9. **Squash:** Plant March to late April.

10. **Tomatoes:** Plant mid-February to March.

Grow Intensive

You can increase the efficiency of your greenhouse by using the growing space inside it wisely. You may find growing spaces that the manufacturer never dreamed of.

One way to save space is to use hanging shelves. Suspend wires from the ceiling and attach two parallel concrete reinforcing rods. Set lightweight plastic flats on top of the rods. Move the rods out of the way when you don't need them.

Common Pest and Disease Problems in the Greenhouse

Even under the best conditions, pests and diseases may bother your greenhouse plants. Good plant health, through good nutrition and environmental control, is the first line of defense. If you find yourself battling pests and diseases, keep them in check by following these tips:

◆ Aphids, mites, and whiteflies can wreak havoc in the greenhouse. Vacuum, squash, and wash away any visible pests at the first sign of trouble. If the problem persists, use soap sprays.

◆ Invite predatory insects by setting up a small "biological island" with pots of parsley-family members, such as dill and chervil, as well as small-flowered ornamentals, such as scented geraniums and salvias. These nectar-providing plants act as hosts to beneficial insects, such as green lacewings, and whitefly parasites, such as *Encarsia formosa*.

◆ Don't run the risk of importing harmful pests and diseases from the garden into the greenhouse by bringing outdoor plants inside. If you must bring those outdoor peppers and eggplants in at the end of the season, quarantine the plants inside sacks made of tightly woven, translucent material for at least 7 to 10 days because it will be easier to detect damage from insects after this time. If problems appear, throw out the plants.

◆ Use preventive sprays to minimize fungal diseases, the greatest disease problems in a greenhouse. Spray fermented diluted seaweed fertilizer on plant leaves at weekly intervals from the seedling stage onward. Isolate or dispose of sick plants.

◆ Keep the greenhouse clean. Isolate or dispose of sick plants, and clean up spilled soil and dropped leaves in the aisles and under benches.

GARDENER TO GARDENER

For Earliest Potatoes, Start Now

I've found that fall-planted potatoes will overwinter wonderfully in my climate beneath thick mulch. So, to save myself some time during the busy spring gardening season, I combine fall potato planting with building new beds. Here's how:

First, I spread a layer of newspaper where I want the new bed to be. I cover the paper with a 1-inch layer of soil or compost and add additional mounds of soil (spaced about 8 inches apart) all across the surface. After planting the potatoes in the mounds, I cover each potato with a little more soil. Finally, I top the whole thing with a thick layer of compost or mulch. After the plants have grown for a while in spring, I simply pull back the mulch or compost and harvest loads of new potatoes. The new garden bed is rich with earthworms and ready to be planted with the next crop.

Sondra Francoeur
Independence, Kansas

Buried Treasures from the Garden

Despite our difficult growing conditions (Zone 2, an altitude of 3,000 feet, and summer temps that go from frying to freezing in 24 hours), our garden supplies us with fresh organic produce year round. Our storage system is the key.

After trying several methods without luck, we finally found the perfect way to keep vegetables through winter. Now, after the first killing frost, we dig up our carrots, beets, parsnips, turnips, and potatoes and pack the unblemished produce into mesh onion bags. We also wrap our cabbages and brussels sprouts in newspaper. Then we dig trenches in the garden, drop in the packed and wrapped veggies, and cover them with a 2-foot-deep blanket of leaves. We mark the area with tall sticks so we can find the veggies easily in snow, then dig them up as we need them throughout the winter.

Even though temperatures can drop as low as -30°F by November, the leafy blanket keeps the vegetables from freezing. Later, snow provides even more insulation. When we uncover the mounds in January, beneath 3 feet of snow, the ground is still soft—and the fresh vegetables are in perfect condition.

Judith Speyers
Prince George, British Columbia

Enjoy Lettuce All Winter

I harvest lettuce right through the winter, even though I garden at 5,000 feet above sea level, where autumn and winter are very cold. Here's what I do:

In early October, I sow lettuce in one of my raised beds, then cover it with clear plastic supported by plastic-pipe hoops. Although the outside temperature drops to 5°F by Thanksgiving, I am able to harvest lettuce from this bed until Christmas. For later winter harvest, I sow lettuce indoors under lights in October. I begin harvesting these greens in December and continue picking for several months.

Douglas George
Loveland, Colorado

GARDENER TO GARDENER

Nifty $25 Greenhouse

To make our greenhouse, I removed the swings from an old swing set, then wrapped the frame with clear 6-mil plastic. I secured the plastic in place with 4-inch-long clamps made from PVC pipe, then mounded soil all around the bottom edges to keep out the cold air. On one end, I installed a zipper door, cut from an old dome tent.

Robert Butcher
Kingston, Oklahoma

Cabbage by the Bucket

In past winters, I had tried to store my harvested cabbage outdoors in the ground, but always had a problem with rodents. Last year, I discovered an easy solution: I simply stored the heads in large plastic buckets covered with leaves.

To keep the cabbage dry, I drilled drainage holes in the bottom of the buckets. I also covered the buckets with screening or ventilated lids before burying them in the leaves. The cabbage remained in good condition throughout winter, and in spring I used the leaves for mulch.

Helen Brink
Greensville, Ontario

Grow Tomatoes Year Round

By growing my tomato plants in large containers year-round, I am able to pick ripe fruits starting in early May! I've kept the same plants going for 3 years now. In summer, the potted plants ('Stupice' and 'Early Girl' varieties) live on my south-facing deck. In fall, I roll the containers indoors, where the plants continue to grow but don't produce much fruit, due to the low light level. Come spring, they're ready to return to the deck for another season of production. As a bonus, slugs and deer are never a problem. I'll never grow tomatoes any other way.

Eugenia Reinauer
Bainbridge Island, Washington

Prepare Now for Spring Planting

To save time in spring, we now begin preparing the soil in fall. We lay heavy cardboard over the garden, cover it with a layer of hay, then add some boards to hold it down. In spring, all we have to do is remove the boards, dig some holes through the decomposing cardboard, and mix in some compost before planting the cukes, melons, and other crops that we've started indoors. Although our garden is quite large, we do no plowing or weeding, and we need to water much less often due to the thick mulch.

Mr. and Mrs. Edward Wanserski
Rosholt, Wisconsin

Blanket Your Soil with Bagged Leaves

I use leaves to keep my ground warm throughout winter, allowing me to plant extra early the following spring.

In fall, I gather as many leaves as possible, putting some onto my compost pile and the rest in plastic trash bags. I lay the bags over areas that I want to plant early next year. When I lift the bags in spring, the ground is soft and ready for sowing cool-weather crops such as kale. And the leaves inside the bags are partially decomposed and ready to layer into the compost pile.

Darleen A. Clements
Seattle, Washington

Chapter 11

November

Putting the Garden to Bed: Cleaning Up, Fall Pruning

Leaves are of the highest value. He who neglects to save them disregards the sources of fertility which nature is kindly offering.

—American Agriculturist *(1864)*

Ready to wrap up the garden for the season? Although the "to-do list" may seem overwhelming, try to think of it as one last chance to be outdoors—surrounded by the sights, sounds, and smells of nature—before the Big Chill arrives. Pick a balmy day, gather your rake and pruning tools, then head for the back 40 (acres, square feet—it doesn't really matter). Got children, grandchildren, or a neighbor's kids you could borrow for the day? Terrific! They can pile leaves while you pull spent plants, clip branches, and mulch beds.

After the kids have reduced the mountain of leaves to a manageable size, turn those leaves into fertilizer and mulch for your garden. Run them through a chipper/shredder, or simply mow them, then dig them into the soil, use them in compost, or spread them over empty beds. (Whatever you do, don't put them on your curbside!)

When you've checked off all (or nearly all) of the items on your list for the day, head indoors for your reward: the first soup of the season, made from your own earthy-sweet carrots, potatoes, and herbs, followed by warm apple crisp. Life is good.

Gardener's To-Do List—November

**If you don't know what USDA hardiness zone you live in,
check the map on page 230 to find out.**

Zone 3

- [ ] Clean up the garden and compost the remains.
- [ ] Cover empty beds with a blanket of compost.
- [ ] Dig up carrots and parsnips before the ground freezes hard.
- [ ] Cover fall-planted pansies and snapdragons with evergreen boughs.
- [ ] Trim broken branches from trees.
- [ ] Rake the orchard floor, then add a fresh layer of mulch beneath young fruit trees and grapes.
- [ ] Put chicken wire collars around the bases of fruit trees to protect them from rodents this winter.

Zone 4

- [ ] Compost spent plants.
- [ ] Dig up root crops before the ground freezes.
- [ ] After the ground freezes, cover perennials with mulch.
- [ ] Pot up some spring bulbs to force into winter bloom.
- [ ] Apply dormant-oil spray to apple and other fruit trees on a mild day.
- [ ] Paint the lower trunks of young trees to prevent winter sunscald.

Zone 5

- [ ] Harvest carrots, brussels sprouts, and cabbage and store them in a cool basement or unheated garage.
- [ ] Clean up the asparagus bed.
- [ ] If the weather cooperates, prepare a few beds now for early planting next spring, then cover them with mulch.
- [ ] Pop spring-blooming bulbs into beds if you haven't done so already.
- [ ] Mulch around pansies and other hardy flowers, but don't smother anything that's still green.
- [ ] Mow fallen leaves, then use them to mulch strawberries and blueberries.

Zone 6

- [ ] Harvest cold-weather-sweetened carrots, brussels sprouts, cabbage, and kale.
- [ ] Continue to thin lettuce and spinach.
- [ ] Hoard a mountain of leaves, then use them to cover beds for early-spring planting.
- [ ] There's still time to divide daisies and to direct-seed Shirley poppies, bachelor's buttons, and larkspur.
- [ ] Set out evergreen shrubs and trees.

- [ ] Weed strawberries and mulch bramble fruits with chopped leaves.
- [ ] Apply dormant-oil spray to fruit trees on a mild day.

Zone 7

- [ ] Start digging up winter carrots as soon as they are big enough.
- [ ] Harvest bunching onions, then plant more in a new site.
- [ ] Plant garlic.
- [ ] Gather blankets for covering lettuce and other half-hardy crops during the first hard freezes.
- [ ] Edge bulb beds with overwintering pansies for a nice look next spring.
- [ ] Trim back faded mums late in the month.
- [ ] Thin larkspur and poppy seedlings, and move bachelor's buttons.
- [ ] Harvest pecans.
- [ ] Use fresh pine needles to mulch strawberries and brambles.

Zone 8

- [ ] Harvest beans, tomatoes, peppers, and other tender crops before the first frost.
- [ ] Begin to harvest fall cabbage, broccoli, and brussels sprouts after cold weather arrives.
- [ ] Plant garlic, and dig and divide multiplier onions.
- [ ] Sow winter cover crops.
- [ ] Gather and bag leaves to use as mulch throughout the rest of the year.
- [ ] Lift caladium corms and store them in damp sand or vermiculite.

- [ ] Set out pansies, dianthus, snapdragons, and ornamental kale.
- [ ] Use fresh pine needles to mulch azaleas.
- [ ] Trim damaged branches from trees.

Zone 9

- [ ] Harvest squash, cucumbers, and your first fall tomatoes.
- [ ] Start planting your winter garden: Set out seedlings of cabbage, celery, and broccoli.
- [ ] Direct-seed peas, beets, carrots, lettuce, Chinese cabbage, spinach, and chard.
- [ ] Renew the herb garden with fresh plants.
- [ ] Dress up flowerbeds with some new starts of alyssum, calendula, and dianthus.
- [ ] Direct-seed bachelor's buttons, cosmos, and hollyhocks.

Zone 10

- [ ] Plant more cool-weather greens, such as lettuce, spinach, and Chinese cabbage.
- [ ] Late this month, prune back a few peppers and eggplants—they'll bear an early crop next spring.
- [ ] Harvest sweet potatoes.
- [ ] Fill empty patio pots with petunias.
- [ ] Plant tropical bulbs, such as freesias and sparaxis.
- [ ] Keep citrus fruits well watered and check often for bird damage.
- [ ] Spray *Bacillus thuringiensis* (BT) on brassicas as soon as you spot leaf-eating caterpillars.

3 Ways to Make Leaves Work for Your Garden

Leaves are among nature's greatest gifts to an organic gardener. Simply gather them up, and before long, soil microorganisms will turn them into a fabulous organic fertilizer or mulch—for free! Use this three-pronged approach to make the most of your autumn riches:

1. Dig them into the soil! Don't wait until spring to replenish your soil's organic matter—it's more efficient to do it now. Simply work a 3- to 6-inch layer of freshly fallen leaves into your vacant garden beds using a tiller or garden fork. (If you use a fork, be sure to shred, mow, or chop the leaves first to speed their decay.) Come spring, those leaves will have broken down into a rich, dark humus—without any help from you.

Composting leaves in your soil is much easier than building a traditional compost pile because you don't need to turn the ingredients or spread the finished compost. And, according to research conducted by Rutgers University, adding leaves to the soil in fall significantly reduces problems caused by soil nematodes—those microscopic worms that feed on plant roots.

2. Mulch with them! Leaves are a terrific free mulch, as anyone who's ever walked through the woods can attest. Simply push aside a layer of that leafy litter and you're sure to see lots of earthworms—a true sign of healthy, organic soil.

Many gardeners apply a 3- to 4-inch layer of shredded leaves to all of their beds every fall. But the benefits of leaf mulch extend far beyond this single season. In winter, leaf mulch protects young perennials from heaving out of the ground during alternate periods of freezing and thawing. And the following summer, the decaying leaf mulch continues to work by controlling weeds and retaining soil moisture.

What's in a Leaf?

Because trees have extensive root systems, they draw up nutrients from deep within the subsoil. Much of this mineral bounty is passed into the leaves, making them a superior garden resource and mulch.

LEAF	% NITROGEN	% PHOSPHOROUS	% POTASSIUM
American beech	0.67	0.10	0.65
Balsam fir	1.25	0.09	0.12
Red maple	0.52	0.09	0.40
Sugar maple	0.67	0.11	0.75
White ash	0.63	0.15	0.54
White oak	0.65	0.13	0.52

In spring, simply use a rubber-tipped rake to loosen and fluff up the leaves. Don't worry if you're a bit late with your spring raking, though. Because the leaves are shredded, they won't mat down the way whole leaves would, so spring bulbs and perennials will be able to push right through their leafy cover.

One reader of *Organic Gardening* magazine is such a leaf mulch fanatic that he collects 30 garbage bags of maple leaves each fall. Half go into his winter compost pile, and half are used as a winter mulch for vegetable and flowerbeds.

For the mulch, he shreds the leaves and spreads a 4-inch layer on top of the beds. He then sprinkles a thin layer of grass clippings over the leaves. The clippings keep the leaves from blowing away and speed their decay. By spring, the mulch is close to finished compost.

3. Compost them! When it comes to improving the soil and helping plants resist disease, nothing compares to compost. A 1-inch layer of compost spread on the surface of the soil prevents plant disease better than any chemical product.

And leaves are one of the few raw ingredients that you can compost alone. In fact, "leaf mold" (leaf-only compost) has been a horticultural mainstay for generations. To make leaf mold, simply gather moist, shredded leaves in large plastic trash bags, or contain them in a wire cage or compost bin. In about a year, you'll have finished leaf compost.

If you want to transform your leaves into finished compost by spring (or sooner), you'll need to add other ingredients. Compost cooks fastest when the ratio of carbon to nitrogen in the pile is 30 to 1—that is, 30 parts of carbon to 1 part of nitrogen. Fall leaves are one of the very best sources of carbon, so you'll need to add nitrogen-rich grass clippings, garden waste, kitchen scraps, or weeds (without seeds).

If you're composting any oak, palm, or eucalyptus leaves—all slow to decay—you'll need to add extra nitrogen-rich material to the pile and turn it often. But, according to an Ohio State University compost researcher, there's at least one big *advantage* to compost made with these slow-decaying leaves: Because it has a fibrous physical structure, it makes an excellent substitute for peat in potting mixes.

No matter what kind of leaves you use, the keys to fast leaf compost are to *shred and wet the ingredients*. If you don't have a chipper/shredder, simply mow over the leaves with a lawn mower. Then, soak them thoroughly with a hose *before* you pile them. (If you water after you've piled, the leaves in the center will stay dry and won't break down.) Alternate layers of the wet leaves with thin layers of a nitrogen material until the pile is 3 to 5 feet high. Turn the pile four or five times over winter. By spring, your disease-fighting, weed-suppressing, soil-feeding mulch will be ready for your garden.

Don't Believe These Leaf Myths!

Don't believe *everything* you hear "over the garden fence"! These popular myths about leaves are absolutely false:

Leaves make soil acidic. According to a soil scientist at the Connecticut Agricultural Experiment Station, this concern is totally unfounded. The truth is, leaf compost can help *neutralize* acid soils, making them closer to the 7.0 pH that many garden plants prefer.

Black walnut leaf compost kills plants. True, black walnut trees exude a compound, juglone, that inhibits the growth of most nearby plants, and that compound is present in black walnut leaves. But one month of composting will destroy all juglone—as well as similar unwanted compounds—in the leaves, assures an Ohio State University compost researcher.

Top 10 Reasons to Use Organic Mulch in Your Garden

By applying an organic mulch to your vegetable garden or flowerbed, you'll save hours of time each year. You should spread a thin layer of wood chips, grass clippings, shredded bark, sawdust, or pine needles on your garden for the following reasons:

1. Mulch prevents most weed seeds from germinating and even makes it easier to pull those that do pop through.

2. It will hold down dirt and keep it from splashing on flowers and vegetables.

3. It will allow you to water less often because mulch keeps the soil cool and moist.

4. It decomposes slowly, releasing nutrients into the soil throughout the process.

5. Mulch encourages beneficial earthworm activity by improving the soil tilth and nutrient content.

6. It will help prevent alternate freezing and thawing of the soil in winter, which can heave plants out of the soil.

7. Mulching helps to prevent diseases by keeping water and soil from splashing onto plant leaves.

8. Compost mulches can help prevent soil erosion. Because soil that contains a lot of humus (finished compost) holds together better, rainwater permeates the soil, moving easily through the spaces between granules rather than running off the soil surface.

9. Most of the ingredients for organic mulches can already be found in your home or backyard.

10. Many types of mulch are attractive additions to the garden site.

Fall Pruning Made Easy

Relax! Pruning doesn't need to be mysterious, deadly serious, or even all that difficult. Before you begin, list your priorities. The main reasons to prune are to keep a plant healthy, to maintain a size or shape that is appropriate for the surroundings, to eliminate hazards, and maybe to get a few more flowers and fruits along the way.

Keep your priorities in mind while you're working, along with the following tips and techniques, and you'll soon feel like a pruning pro!

What to Cut

Here's what to prune in order of importance:

◆ All dead and dying branches

◆ Diseased limbs

◆ Branches that obstruct views or traffic (pedestrian or otherwise!)

◆ Branches that cross each other (prune the smaller of the two branches)

◆ Branches that are not in symmetry with the plant's natural shape

Where to Cut

That's easy—always cut just beyond the *branch collar* (the raised area where a branch meets the trunk) without cutting into the collar itself.

If you have a hard time locating a specific collar, look for a distinct collar on another branch as a cue to how far from the trunk you should cut. Be brave and check these cuts in about a year. If you've pruned correctly, a circle of healthy callus that looks like a doughnut will have formed around the cut. For bigger limbs that you have to cut with a saw, see the illustration on this page.

Never remove more than one-third of the branches of a tree or shrub in any one pruning session, advises the National Arborist Association in Amherst, New Hampshire. The same rule applies to each individual tree limb that you want to keep: Never remove more than one-third of it, or you might not leave enough foliage to support its future growth. Always avoid creating too much stress on the plant (and yourself!).

Choose the Right Tool

All you really need for painless pruning are a few good tools. Here's what to look for:

Hand pruners. Opt for bypass hand pruners over anvil types, which tend to crush fragile stems. Ergonomic models have handles set at an angle to the blades so you don't have to bend your wrist as much.

Pruning saws. Look for a curved pruning saw that cuts on both the push and pull strokes; it will work more quickly. Try out models in your local hardware store or garden center until you find one that feels and cuts comfortably.

Use a *drop cut* to remove large tree limbs safely. Here's how:
1. Make an undercut about one-third to one-half of the way through the branch.
2. Make a second cut into the top of the limb, just outside the first cut (most of the branch will then snap off at that spot).
3. Make the final cut just outside of the branch collar.

Which Plants to Cut

What plants should you prune now? Plenty! Sharpen your shears and saw, then turn your attention to these:

Summer-flowering shrubs. Prune summer-flowering shrubs anytime from late November to early spring, before their spring growth begins. As a rule of thumb, prune them right after flowering so that you don't cut off the buds for next year's blooms. (That means now is *not* the time to prune spring-flowering shrubs, such as lilacs, redbud, forsythia, azaleas, spirea, or viburnum. If you do, you won't see any flowers next spring.)

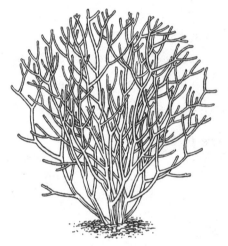

When pruning an overgrown shrub (*left*), avoid using **heading cuts** (*right*) to shorten branches. Besides creating a silly-shaped plant, heading cuts stimulate new growth that is easily killed by cold temperatures.

Instead, use **thinning cuts** (*left*) to remove branches all the way back to their point of origin. The result: a nicely shaped, healthy plant.

Vines. Most vines require little care other than pruning to limit their growth. If you decide to prune your vines, late fall (after a few hard freezes) is a good time to clip away the excess growth of English ivy, Boston ivy, Virginia creeper, and other vines on building walls and around windows, says Donald A. Rakow, Ph.D., landscape horticulturist at Cornell University.

Herbaceous perennials. Cut back the foliage of most herbaceous perennials to spruce up the garden and reduce the chances of disease or insects overwintering in dead leaves and stems. (Exceptions are evergreens, such as sea pinks, European ginger, evergreen candytuft, *Sedum spurium*, and some ferns.)

Dead and diseased plants. Fall is the perfect time to prune dead or diseased wood from trees and shrubs, before winter wind, ice, and snow prune them for you! Large falling branches can cause major damage or injury, so don't delay.

When pruning trees and shrubs that have wilt, canker, fire blight, gall, or other diseases on a portion of the plant, take care not to spread the disease to the plant's healthy areas. Don't prune when the plants are wet (water helps spread disease), and always cut well below diseased areas—as much as 6 inches—to ensure you get all of the infected wood. Experts now agree that it is not necessary to sterilize your equipment as long as you cut well below the diseased area. Pruning cuts, even on large branches, need no paint or other wound dressings applied. They will heal faster on their own.

Warnings

Some plants should not be pruned in fall, when they are going into dormancy. If you prune these sensitive species now, they'll be more susceptible to winter damage and insect infestation. Wait until late winter or early spring to prune:

♦ Deciduous trees that are "borderline" hardy in your area

♦ Canker-prone species (such as crabapple, Callory and Bradford pears, cherries, plums, honey locust, maple, spruce, willow, and poplar)

♦ Conifers (trees with needles)

♦ Fruit trees and bushes

Major pruning jobs—such as rejuvenating shrubs through severe pruning—usually are better left for early spring, as well.

Finally, think about your safety and the safety of buildings or plants around the targeted plant. If you have any doubts about your ability to prune something (especially large trees) safely, call in a professional.

Don't Haul Those Twigs Away!

A pile of branches isn't a liability that you have to spend hours chipping or dragging to the dump. If your property has a secluded corner, stash your branches there and create an instant wildlife shelter. Birds, squirrels, rabbits, and other creatures will take advantage of your generosity, and you'll gain the pleasure of their company.

Pruning and Training Glossary

If you're ready to dust off the pruning saw and head for an overgrown shrub, you may want to brush up on the lingo that goes along with pruning and training.

Branch collar \\'branch 'kä-lər\\:
The bulge at the base of a branch; this part of the trunk helps hold the branch to the trunk.

Branch crotch \\'branch 'krä-ch\\:
The angle at which the tree branch meets the trunk or parent stem.

Break bud \\'brāk 'bəd\\:
When a latent bud is stimulated into growing out into a leaf or twig.

Cane \\'kān\\:
A long and slender branch usually originating from the roots.

Heading cut \\'he-diŋ 'kət\\:
To cut back a branch to a side bud or shoot.

Leader \\'lē-dər\\:
The main, primary, or tallest shoot of a trunk. Trees can be single-leadered, such as birch, or multiple-leadered, such as vine maple.

New wood \\'nü 'wùd\\:
A cane of the current year's growth. Some shrubs bloom only on new wood, while others bloom on the previous year's growth.

Old wood \\'ōld 'wùd\\:
A cane of the previous year's growth or cane that's older.

Pinching \\'pinch-iŋ\\:
To use your fingertips to squeeze off the end bud of a twig or stem to make the plant more compact and bushy.

Skirting or limbing up \\'skər-tiŋ, 'lim-iŋ, 'əp\\:
To prune the lower limbs of a tree to increase air circulation, improve visibility, or clear room underneath the tree.

Sucker \\'sə-kər\\:
An upright shoot growing from a root or graft union. In common usage, these are straight, rapid-growing shoots or watersprouts that grow in response to poor pruning and wounding.

Thin out \\'thin 'aút\\:
To cut off a limb at the base, either at ground level or at a branch collar.

Topiary \\'tō-pē-ˌer-ē\\:
Plants sculpted into geometric shapes or likenesses of animals or people.

Technique

Fill Your Home with Blooms from Bulbs

Here's a fun project that you can do now to lift your spirits later this winter. Pot up some spring-blooming bulbs, chill them outdoors for a few months, then bring them inside to add fragrance and color to your home, just when you need them most.

This low-tech forcing technique eliminates the need to chill bulbs in a refrigerator. After about 12 to 16 weeks of outdoor chilling, plug the individually rooted bulbs into larger containers, a window box, or wherever you like. After they've finished blooming, plant the bulbs in your garden.

You can force just about any daffodil, tulip, or hyacinth variety this way, but short early-blooming types are easiest.

1. Fill individual pots (at least 3 inches wide × 4 inches deep) with potting soil, leaving an inch of space at the top. Place one bulb on top of the soil inside each pot (don't bury the bulbs). Pack enough additional potting soil around the base of the bulb to hold it in place while it forms roots.

2. Put the pots into a nursery flat or tray, then set the whole thing outdoors in a shaded location. Cover the pots with a thick layer of leaves or pine needles (at least 6 to 8 inches). If you've had problems with critters digging up your bulbs in the past, cover the pots with metal screening before you mulch them.

3. Leave the bulbs outdoors for at least 12 to 16 weeks. Check the potting mix to be sure that it stays moist. When you see roots starting to emerge from the bottom of the pot (near the end of the chilling period), the bulbs are ready to repot into larger containers. This time, plant the rooted bulbs at the normal depth.

To chill bulbs outdoors, plant them shallowly in individual pots. Cover the pots with a thick layer of mulch.

Easy Bulbs for Forcing

These spring-blooming bulbs are good choices for forcing:

DAFFODILS
'February Gold'
'Jack Snipe'
'Peeping Tom'
'Tete-a-Tete'

HYACINTHS
'Delft Blue'
'Pink Pearl'

TULIPS
'Christmas Marvel'
'Flair'

GARDENER TO GARDENER

Turn Faded Perennials into Mulch

Instead of cutting back my spent perennials in autumn, I leave the faded foliage in place as an added mulch to protect the crowns when there is no snow. When we do get snow, it tends to collect around the foliage first, adding to its insulation value. Additionally, the spent foliage provides winter interest and serves as markers so that I know where to (and where not to) dig in spring.

Mark Yantek
Cleveland, Ohio

Shield Bulbs from Critters

I garden on 2 acres of pure sand, teeming with moles, voles, gophers, and mice. To keep these underground critters from eating or displacing my bulbs, I make "bulb barriers" from those gallon-size plastic pots that perennials and shrubs are sold in. (I get dozens of the containers each spring when I buy new plants, and this is a great way to recycle them!)

To make a bulb barrier, use a linoleum knife to cut off the bottom of a container. (You can leave the bottom on if you're feeling lazy, but removing it leaves more room for root growth.) Then dig a hole wide enough to accommodate several containers and deep enough so that their top rims are about 2 inches below the surface of the soil.

After the containers are in place, toss a handful of bonemeal and about an inch of soil into the bottom of each and follow with three large (or four medium-size) bulbs. Fill the pots with the soil you removed from the hole, then water the area. The bulbs will do fine in these pots for about 6 or 7 years before becoming overcrowded. When I notice that the crowding is causing the blooms to decrease in size, I dig up the pots, divide the bulbs, and replant.

Karen Hampton
Oklahoma City, Oklahoma

Box, Don't Bag, Your Leaves

Don't wrestle with raked leaves to get them into plastic garbage bags. Instead, lay a large cardboard box on its side and rake your leaves into it. It's much easier than using leaf bags—especially if you're working by yourself. If you don't already have boxes on hand, check with your local grocer. Supermarkets often receive large boxes for toilet paper and other products sold in bulk. These boxes hold a lot of leaves. And even when they're full, they're light enough to drag easily across the lawn.

Della Kapocius
Grand Forks, North Dakota

GARDENER TO GARDENER

Grow a Garden in Leaves Alone

Some gardeners look at autumn leaves as a chore, but for me, they're pure treasure! For years, I've been growing vegetables, flowers, and herbs in beds that contain only autumn leaves (and a little garden soil), stuffed into a wire frame. The plants do fine, and the beds are almost maintenance free because weeds don't grow in the leaves that surround the plants. After a couple of years, the leaves break down into rich, dark humus that I then add to my regular garden beds.

To do this yourself, you'll need only lots of leaves, some kind of wire frame, and a few stakes to support the wire. Chicken wire is OK, but I prefer the heftier, more-durable wires. Make the frames any size or shape you like—mine are 30 to 36 inches high.

Now, get lots and lots of leaves. I've packed 90 bags of leaves into a 4 × 8 × 3-foot leaf bed, and could have used more! I never have enough leaves of my own, so I beg my neighbors for their bagged leaves and cruise the streets on garbage night.

The reason you need so many leaves is that you must pack them down tight. Start filling the frame with leaves at the beginning of fall, and add more as the season continues. After the frame is filled, top the bed with 2 or 3 inches of soil. To keep the soil from spilling outside the wire, tuck a barrier of grass clippings or straw around the inside edges of the frame before adding the soil.

If you can't spare enough garden soil to cover an entire leaf bed, make planting holes in the leaves and fill these with 3 or 4 quarts of soil apiece.

In spring, plant your seedlings in the soil. Be sure to keep the beds watered, especially when the seeds are germinating or the seedlings are getting established. (Even a tightly packed leaf bed is relatively porous and will dry out more quickly than a regular bed.) As the plants grow, the roots will grow downward into the decomposing leaves to obtain all the nutrients they need. I've never had to do any supplemental feeding.

Don't worry when you see the beds' height shrinking due to the decaying leaves—it doesn't bother the plants one bit! After a couple of seasons, each leaf bed will have shrunk down considerably.

Patricia Leuchtman
Charlemont, Massachusetts

Winter Protection for Roses

Prepare roses for the onset of cold weather by recycling plastic nursery pots (1 gallon or larger) into protective collars. Using a utility knife, cut out the bottom and up one side of each pot, so that it opens up, like the letter C. Then, simply wrap a pair of the cut pots around each rose bush to make a collar. Fill the inside of the pots with shredded leaves and compost. In spring, remove the collars and spread the mulch around the bases of the roses.

Mary Leunissen
Guelph, Ontario

Recycle Your Okra Stalks

I've discovered that spent okra stalks make a convenient natural trellis for peas and other climbing crops. I leave my okra stalks in the garden over the winter, then plant peas at the base of the stalks in early spring. When the peas are finished, I pull up the whole works and have a ready row in which to plant tomatoes, peppers, or anything I missed in the spring rush.

Dolores Bainum
Camden, Tennessee

GARDENER TO GARDENER

Snow Shovel Does Double Duty

In autumn, we wait for all of the leaves to fall from the trees before we gather them. Then, we simply push them into one big pile, using a plastic snow shovel. Because the shovel is plastic, the edge slides over the lawn or pavement, scooping up everything in its path without causing any damage to the grass.

Sue Blyth
Ottawa, Canada

Turn Leaves into Easy Mower-Made Mulch

Chopped leaves make an ideal mulch for a wildflower garden—especially a woodland one. If you have a side-discharge mower, try this technique to chop and spread leaves while you mow.

Start mowing your leaf-covered yard on the opposite side from and parallel to the garden bed you want to mulch. Keep the discharge pointed toward the garden with each pass, moving gradually toward the garden. You'll create a moving windrow of chopped leaves. As you finally mow along the garden's edge, the last of the shredded leaves will be blown into place.

If your mower discharges into a bag, just mow your leaf-covered lawn and then empty the bag directly onto your beds.

Christine Esposito
Raritan, New Jersey

Garden Cart Retrofit

My garden cart works fine for most of my garden chores, but when it comes to transporting leaves in fall, its 16-inch-high sides fall short. To raise the height of the sides another 16 to 24 inches (so that I can haul really big piles of leaves), I slide old window screens between the leaves and the sides of the cart. The cart now has double the leaf capacity or more, meaning fewer trips and better use of my time and energy.

Elliott Berenson
Chesterland, Ohio

Chapter 12

December

Wildlife in the Garden: Bringing In the Birds, Keeping Out the "Pests"

In drear nighted December
Too happy, happy tree
Thy branches ne'er remember
Their green felicity.

—*John Keats*

Come December, gardeners in most regions view their landscapes from a different perspective. Without the green of growing plants to capture our attention, other elements of the natural world come into sharp focus. Wildlife is easier to observe—providing us with hours of entertainment . . . or frustration.

We gardeners have a love-hate relationship with wildlife. As long as the deer, squirrels, raccoons, and birds respect "our" boundaries, they remain loveable, untamed pets. But when Bambi or Rocky crosses the line that defines our growing space, a mysterious transformation seems to occur: Those untamed pets become monstrous pests!

Maybe it's time *we* change. (After all, who is the real trespasser here?) Some wildlife is truly beneficial to gardens and prey on common garden pests, making control products unnecessary. Other animals are more problematic. But ingenious gardeners, as you'll soon read, have devised all sorts of benign, yet effective, deterrents.

Gardener's To-Do List—December

**If you don't know what USDA hardiness zone you live in,
check the map on page 230 to find out.**

Zone 3

- [ ] Pot up a few more bulbs to force into bloom for early spring, and move those that already have sprouted into bright light.
- [ ] Check stored vegetables for signs of spoilage.
- [ ] Wrap the bases of fruit trees with wire mesh to protect them from hungry rodents.

Zone 4

- [ ] After the soil freezes, mulch over the crowns of perennials to keep them from heaving out of the ground during winter thaws.
- [ ] Pot up some spring bulbs to force into bloom; keep them cold (but not freezing) until they sprout.
- [ ] If whiteflies and mealybugs are attacking your houseplants, set the plants in the shower and wash your troubles down the drain.
- [ ] Inspect stored fruits, bulbs, and corms for signs of spoilage.
- [ ] Check the bases of tree trunks for mouse marks; encircle the trunks with wire mesh to prevent further damage.

Zone 5

- [ ] If you've run out of straw to cover carrots, parsnips, and salsify in the garden, dig up these root crops and store them in a cool basement.
- [ ] Harvest brussels sprouts and cabbage plants, roots and all; they'll keep for weeks that way if stored in a cool basement or root cellar.
- [ ] Cover parsley with milk jug cloches, then surround the covers with insulating leaf mulch.
- [ ] Protect overwintering spinach with row covers.
- [ ] Wrap chicken wire cages around young fruit trees to protect them from deer.
- [ ] Before the ground freezes, dig planting holes for any trees you intend to plant in late winter.

Zone 6

- [ ] Cover perennial flowers, as well as pansies and snapdragons, with evergreen boughs to protect them from ice and harsh winds.
- [ ] Harvest brussels sprouts and cabbage.
- [ ] Mulch empty vegetable beds with chopped leaves.
- [ ] To enjoy spinach throughout winter, cover the plants now with a plastic tunnel.

- [ ] Rake up all fallen apples and apple leaves to stop the spread of scab and other orchard diseases.
- [ ] On a mild day, apply dormant-oil spray to smother scale and other sap suckers.

Zone 7

- [ ] Harvest brussels sprouts, kale, cabbage, and collards.
- [ ] Mulch Jerusalem artichokes, carrots, parsnips, and other crops that will spend winter underground.
- [ ] Spread mulch over beds where early spring crops will grow.
- [ ] Turn compost one last time, then cover it with a tarp to prevent nutrients from leaching away during winter rains.
- [ ] Dig, divide, and replant crowded bulbs.
- [ ] Continue setting out hardy annual and perennial seedlings, then cover them with cloches.
- [ ] On a mild day, apply dormant-oil spray to smother scale and aphids.

Zone 8

- [ ] Harvest frost-sweetened spinach, kale, and collards.
- [ ] Pile compost and leaf mulch on vacant garden beds.
- [ ] Near the end of the month, start seeds of cabbage and hardy lettuces indoors.
- [ ] Plant spring-blooming bulbs.

- [ ] Direct-seed sweet peas, larkspur, Shirley poppies, and bachelor's buttons.
- [ ] Continue planting pansies, dianthus, and snapdragons.
- [ ] Prune roses, then replace their old mulch with oat straw to discourage black spot.

Zone 9

- [ ] Plant cool-season vegetables, such as lettuce and cabbage-family crops. Be patient—they grow slowly during winter's short days.
- [ ] Around month's end, start seeds of tomatoes, peppers, and eggplant indoors.
- [ ] Direct-seed poppies, bachelor's buttons, nigella, cosmos, and larkspur in well-drained beds.
- [ ] Apply a fresh layer of clean mulch to berries and grapes.

Zone 10

- [ ] Plant all sorts of greens, beets, carrots, and peas.
- [ ] Around month's end, start seeds of tomatoes, peppers, and eggplant indoors.
- [ ] Pop bedding plants, such as pansies, dianthus, and snapdragons, into both containers and empty spots in beds.
- [ ] Plant ornamental shrubs and trees.
- [ ] Enjoy your homegrown citrus fruits.

A Gardener's Best Friends

The best garden aid you could ever invest in is probably sitting outside your window right now, ready to work for peanuts—or sunflower seeds. See that chickadee out there—the one with the bright black eyes and friendly cocked head? No doubt he's one of your feeder favorites. Now take a look at those starlings lurking nearby, ready to drop down by the dozen, scaring off your small feeder guests. You've probably wondered what you could do to discourage them. Big mistake. Every bird in your yard is valuable, especially in winter. That's when future populations of pests get stopped dead— literally—by the sharp eyes and quick beaks of the feathered army.

No Such Thing as Eating Like a Bird

Watch a bird when it leaves your feeder and you'll see that it constantly is searching out bits of food from every part of your garden. Finches sway on the stems of lamb's-quarter and ragweed, devouring seeds that would otherwise sprout in your beds. Sparrows, juncos, and towhees scratch beneath plants, vacuuming up fallen seeds and any insects that cross their path. Even the common house sparrow helps out, eating thousands of seeds of crabgrass, chickweed, purslane, and other weeds. Chickadees tear open cocoons of bagworms and other pests, munching the tasty morsels inside.

Nuthatches scour trees top to bottom in a comical upside-down posture that gives them a great vantage point for picking insects and their eggs out of cracks in the bark. The brown creeper covers niches that nuthatches overlook, spiraling from bottom to top of the same trees. Woodpeckers whack their way into tree bark and plant stems, ferreting out hidden larvae and harmful ants and beetles. Even those scorned starlings earn their stripes, patrolling your lawn with sharp, stabbing beaks for Japanese beetle grubs.

Consider that a single chickadee—maybe the very one who's entertaining you outside the window right now—can down a thousand or more scale insects, aphids, codling moths, pear psyllas, pine weevils, and other delicacies in just one day, and you get some idea of the immensely positive potential of a bird's appetite. That's why it makes such great gardening sense to invite more birds to your yard. Fewer pests mean healthier flower gardens, bigger and better vegetables, long-lived trees and shrubs, and less work for you.

In a single day, one diminutive house wren can consume 500 beetles, grubs, and insect eggs. More than half of a chickadee's winter diet is aphid eggs. And a swallow can consume incredible numbers of flying insects—by one researcher's count, a single swallow ate more than 1,000 leafhoppers in less than a day.

A Delectable Landscape

Unless your property is barren of trees, shrubs, and flowers, you'll be delighted with bird visitors who regularly stop by to check out your offerings. But you can attract far more birds by offering a few bird amenities, such as berry-producing shrubs and fruit trees, water for drinking and bathing, and nesting spots in trees and birdhouses. Your first visitors will undoubtedly be robins, nuthatches, hummingbirds, titmice, cardinals, and sparrows, but with a little extra enticement, you can welcome special birds, such as bluebirds, goldfinches, and juncos.

Feeders, birdbaths, and birdhouses play a vital part in bringing birds to your backyard garden. But trees, shrubs, and flowering plants can fill the same role without extra effort from you! Plants provide food, cover, and nesting sites, and because they trap dew and rain and control runoff, they help provide water, too.

When choosing plants, look for food-bearing species that will provide fruits, buds, and seeds throughout the year. Mix deciduous and evergreen varieties in order to provide cover and shelter all year-round. Native species are usually the biggest bird attractant, and local birds will turn to them first for food and cover.

Doing Your Part

In winter, a feeding station is the best way to lure the anti-pest squad to your yard. It takes a lot of fuel to keep those hyperactive bodies humming, so easy pickings are always welcome. But don't worry about your feeder birds getting lazy. Maybe bugs and weed seeds taste better than birdseed, or maybe it's just in the genes, but birds will still glean your garden no matter how delicious a banquet you spread.

For results right now, all you need are a few dollars' worth of seed, a simple cylindrical or open tray feeder, and a plastic mesh onion bag stuffed with suet (or freebie beef trimmings). If you like, add a source of water or a discarded Christmas tree for shelter to make your yard an even better place for birds to linger. Within a week or two, you'll have a host of garden helpers for pest control at its finest. You'll also enjoy the pleasure of their company. When your garden is alive with birds, even the grayest winter day can be a delight.

Holiday Feast

Winter birds love a Christmas feast even after the holiday. Set your Christmas tree outside where it's sheltered from the wind, then gather leftovers from holiday baking to make garlands of raisins, nuts, and dried fruit slices. For "ornaments," fill citrus rinds with peanut butter and fruit.

A simple cylindrical feeder and a source of water will keep your backyard birds happy throughout winter.

GARDENER TO GARDENER

Benefits of Bayberries

If you're a bird lover, be sure to try planting bayberries (Myrica pensylvanica). This easy-to-grow shrub offers not only evergreen shelter year-round but also a winter banquet for birds. More than 20 kinds of birds eat the berries, including pheasants, wild turkeys, flickers, downy woodpeckers, chickadees, Carolina wrens, yellow-rumped warblers, and tree swallows. The native shrub thrives in many conditions, including poor, sandy soils. Plant it in full sun to partial shade in Zones 2 to 7.

*Mike Hradel
Coldstream Farm,
Michigan*

Winter Berries for the Birds

Bird watching can be habit forming. The chair by the bird feeder window is often the most popular seat in the house. But not all birds live on feeder meals alone. Some simply will not eat from a feeder, no matter how tempting the treats are. To attract the largest variety of birds into your backyard in the cold months, you must also plant the right shrubs, bushes, and trees for them.

The best bets are those plants that produce plenty of berries and hold them on their branches into winter, when other food sources become sparse. These plantings can provide not only food but also shelter for your wild, feathered visitors all year long. Additionally, many of these bushes can serve as foundation plantings and bring birds close to your viewing window.

Birds are gluttons when it comes to berries: They can't eat just one. Flocks of robins or tanagers will return day after day until every berry on your dogwood is history. And once the berries are gone, so are the birds.

You can extend the season by planting different cultivars and types of berries. A combination planting of shining sumac, elderberry, and winterberry holly, for example, will ensure a steady level of bird activity around your yard.

Going Native

When choosing trees and shrubs for the birds, always try to use native varieties no matter where you plant. Natives will thrive naturally in your soil and climate. And you rarely need to fuss with soil improvements, fertilizer, or extra watering.

Also, research shows that foraging birds seek out the familiar over the exotic species. Putting native plants into your garden and yard will give you an edge in attracting birds. So instead of planting a hedge of forsythia, which has little bird value, try a food-providing hedge of bayberry, blueberry, elderberry, or sumac. It can work as a border planting or hedgerow, and will supply tasty morsels for many birds.

Birds don't care how good your landscape design skills are, but a yard with bird appeal should also look good to your eyes. Whether you like straight edges, right angles, and controlled plantings of a formal style, or the casual plantings and curving lines of an informal or naturalistic garden, you can take steps to build bird appeal. Remember that birds are more likely to take up residence in an area that has little traffic. If you can keep at least part of your yard undisturbed, or nearly so, with low-maintenance or naturalistic plantings, birds will soon build nests in the hedges, shrubs, and trees.

There are many attractive berry bushes for the birds, but here are four genera—dogwood, holly, sumac, and viburnum—whose varieties and cultivars are adaptable to a wide range of soil conditions and climates. (For additional choices, see "More Trees and Shrubs to Attract Birds" on page 203.)

Dogwoods

Dogwoods (*Cornus* spp.) are always a big hit with fall- and winter-feeding birds. Many of these birds are accustomed to seeking out dogwoods because so many of the native species bear fruit. Bluebirds, grosbeaks, tanagers, thrushes, vireos, waxwings, woodpeckers, and a host of other birds adore the dogwoods' red, shiny berries.

Most of the North American varieties are shrubs with casual growth habits. Flowering dogwood (*Cornus florida*) is a superb small tree for a partly shaded site to Zone 5. Other, lesser-known dogwoods are just as popular with birds, although they seem to be a well-kept secret among gardeners. Many species are available, including pale, or silky, dogwood (*C. obliqua*) and red osier dogwood (*C. sericea*), two shrubs that make fine windbreaks or cover habitat at feeding stations; they bear clusters of stemmed fruits that are eagerly stripped by birds. The native gray dogwood (*C. racemosa*) has attractive blue-green foliage, white flowers, and whitish berries, and it makes an excellent shrub for either a landscape border or naturalized area.

Native dogwoods may be sold at extremely reasonable prices by your state department of natural resources or through county extension offices. You also can check with a nearby nature center or native-plant specialist to get recommendations for your region.

Hollies

When holly berries are ripe, every day is Christmas for the birds. Many varieties of birds will come and visit continuously until all the berries are gone. The berry-laden branches of the deciduous variety 'Winter Red' can be stripped bare in just minutes! Hollies attract at least 12 species which eat its berries, including cedar waxwings, eastern bluebirds, and northern mockingbirds.

Most of us think "evergreen" when we think of hollies, but deciduous types hold great appeal for birds, too—especially for bluebirds, which seem to find holly berries irresistible. Unlike evergreen hollies, some of which can grow to 100 feet tall, deciduous hollies (*Ilex verticillata* and hybrids) are shrubby plants that grow only about 6 to 9 feet tall. 'Winter Red' produces huge amounts of gorgeous, bright red berries on 9-foot-tall plants. If you have a small yard, choose 'Red Sprite,' which grows to just 3 to 5 feet in height.

Water in Winter

Freezing temperatures and a lack of food aren't the only problems that birds face in winter. When creeks and ponds are locked in ice, birds can go thirsty.

Consider equipping your birdbath with an electric or solar deicer to keep water available to birds all winter. To attract even more birds, add a recirculating pump—birds will flock to the sound of running water.

Include These Vines in Your Wildlife Garden

Vines make a perfect centerpiece for a garden designed to attract birds, butterflies, and other wildlife. Hummingbirds are drawn to colorful blooming vines with tubular flowers, such as cypress vine (*Ipomoea quamoclit*), trumpet honeysuckle (*Lonicera sempervirens*), and orange trumpet vine (*Campsis radicans*). Butterflies are attracted to cypress vine and sweet peas (*Lathyrus latifolius*).

All of the hollies are perfect for planting in groups or in a mixed border outside the window of your favorite easy chair. Hollies are pollinated by insects rather than wind, so be sure to plant at least one male tree nearby. A nursery center employee may be able to identify the sex of the tree before you buy. Evergreen hollies are hardy to Zone 5; deciduous types are even tougher, thriving as far north as Zone 3.

Sumacs

Sumacs (*Rhus* spp.) get short shrift in most home landscapes, probably because they are such common wild plants. They become part of the backyard scenery, hardly drawing a second glance.

Only in autumn, when sumac leaves become burnished with crimson or orange, and in winter, when their branches stand bare and stark with a candelabra of deep red fruits, can sumacs be considered eye-catching and landscape worthy.

Sumacs flourish from Zones 2 to 10, depending on the species, and in an assortment of conditions. Most grow best in full sun to light shade and are very drought tolerant.

Winter is when sumacs become standouts with birds. Although sumac berries aren't a preferred food, as are acorns or nuts, they are very much appreciated by birds as a meal of last resort. Those spires of fuzzy fruits provide long-term food in the leanest months of the year, when little else is available. Some of the best winter bird watching often can be had from spots that overlook a stand of gangly sumacs.

Viburnums

Viburnums are a diverse group of more than 150 shrubs and small trees. They bear plentiful blooms, and some have fall foliage that's as brilliant as that of a sugar maple. They are fast growing, too, making them fine for a hedge or shrub grouping.

Viburnums differ greatly in hardiness, although most adapt to a range of soil types, moisture levels, and light conditions. Two American natives, nannyberry viburnum (*Viburnum lentago*) and downy arrowwood (*V. dentatum*), are among those that flourish as far north as Zone 2. Others, including many Chinese species and cultivars, are far more tender, surviving only to Zone 9. In general, however, most species are hardy to at least Zone 6.

The berries come in red, blue, and orange, as well as the blue-black of nannyberry and arrowwood. Plant breeders have developed dozens of garden-worthy viburnum cultivars, but the berries of some have lost their appeal to birds. If you want surefire berries for birds, stick to native species, which will thrive naturally in your garden.

More Trees and Shrubs to Attract Birds

These lesser-known landscape plants also are excellent for attracting birds to your yard in winter. Check mail-order catalogs that specialize in native plants.

Manzanitas (*Arctostaphlyos* spp.): Striking, smooth red bark characterizes many of the shrubby species of this western and southwestern genus, which also includes ground-hugging bearberry (*A. uva-ursi*). The red or brown berries are prime food for grouse, grosbeaks, jays, and mockingbirds. (Zones 2 to 6)

Hackberries (*Celtis laevigata*, *C. occidentalis*, and other species): These lovely shade trees have interesting warty gray bark and a multitude of small fruits that ripen in late fall to early winter. Hackberries are a source of food for flickers, jays, mockingbirds, orioles, sapsuckers, thrashers, thrushes, titmice, woodpeckers, and wrens. (Zones 5 to 9 for *C. laevigata*; Zones 2 to 9 for *C. occidentalis*)

California dogwood (*Cornus californica*): This dogwood is a deciduous western native shrub with dark red bark and white berries. It is considered a hybrid of western dogwood (*C. occidentalis*) and red osier dogwood (*C. sericea*) by some taxonomists. Its berries are eaten by songbirds, crows, grouse, quail, and partridges. (Zones 4 to 8)

Persimmons (*Diospyros virginiana*, *D. texana*): Persimmons are native trees with large, simple leaves and bark that's checkered like an alligator hide. The astringent orange or orange-red fruits are fleshy and turn soft and sweet when ripe. They are beloved by bluebirds, catbirds, mockingbirds, robins, sapsuckers, starlings, and waxwings. (Zones 4 to 9 for *D. virginiana*; Zones 7 to 9 for *D. texana*)

Greenbriers (*Smilax* spp.): These evergreen or deciduous native climbers, some thorny stemmed, are absent in parts of the West but abundant in the southeastern United States. Berries may be yellow, black, blue, or green and are devoured eagerly by many birds. Bluebirds, catbirds, fish crows, flickers, mockingbirds, fox sparrows, white-throated sparrows, thrushes, and pileated woodpeckers are among the prime customers. (Zones 3 to 8)

American cranberry (*Viburnum trilobum*): An eye-catcher in fall, this native bears heavy crops of glowing red fruits. The upright shrub grows to 9 feet in height. Its white spring flowers may have some fragrance; fall color can be striking or muted, in shades of yellow to red. The fruits are attractive to eastern bluebirds, cedar waxwings, bobolinks, cardinals, and mockingbirds. (Zones 2 to 8)

Bring In the Birds

Besides providing food for birds this winter, try these additional tips for making your yard a year-round haven for them:

♦ **Offer water.** Provide a birdbath or other source of drinking water.

♦ **Make a home.** To attract wrens, hang a house with a 1-inch-diameter hole from a branch in a fruit tree 5 to 10 feet above the ground. Bluebirds prefer a house with a 1⅜-inch-diameter hole mounted on a post 4 to 6 feet above the ground. Robins and swallows prefer an open nesting shelf 6 to 10 feet above the ground.

♦ **Plant a hedgerow.** A naturalistic hedgerow will attract those nesting birds that don't use bird boxes.

Help the Birds Get a Grip

When building a birdhouse, choose rough-sawed boards rather than finished lumber to give little birds a better grip on the roof, sides, and front of the house. Rough lumber also happens to be less expensive!

Technique

Squirrel-Proof Your Birdhouse

You may have visions of happy bird families when you mount that new birdhouse this spring, but your neighborhood squirrels have other plans. Rambunctious squirrels often gnaw the entrance holes to enlarge them so that they can move in instead of the birds. Many squirrels will even chew and enlarge the feeder holes of wooden feeders in the hopes of accessing more seed.

There are a couple of ways to prevent squirrels from invading birdhouses. Your first line of defense is a baffle. Baffles are metal or plastic guards that are mounted or hung on the pole between the birdhouse and the ground, so that climbing animals can't launch a sneak attack. Squirrels are determined though, and they may eventually maneuver around a baffle. You can also squirrel-proof nest boxes with a slate or metal barrier to keep out the varmints. The guards are easy to cut and install, although you may find it easier to have a masonry or metal shop make them for you.

You will need a square of slate or sheet metal that's at least twice the size of the birdhouse's entrance hole. For example, if the entrance hole of the birdhouse measures 2¼ inches in diameter, the slate or metal square should be no less than 4½ inches square.

Drill a hole the same size as the entrance hole in the center of the slate or metal square, using a masonry drill bit if drilling through slate. Place the guard flush with the front of the birdhouse, aligning the holes. Drill into the wood along the outer edges of the guard, then use screws to attach the guard to the birdhouse.

For ready-made birdhouses fitted with antisquirrel devices, contact Coveside Conservation Products, 202 U.S. Route 1-PMB 374, Falmouth, ME 04105; (207) 774-7606; fax (207) 774-7613; www.maine.com/coveside.

A metal or slate guard will keep pesky squirrels from using their teeth to enlarge the entrance hole to suit their own bodies.

Wintertime Treats

Make your backyard a haven for birds by supplying them with suet and other treats. Suet provides loads of calories to fuel the high-speed metabolism of a wide variety of feathered friends. Chickadees, jays, titmice, woodpeckers, and nuthatches, just to name a few, will flock to your wintertime offerings.

Suet Cakes

    1 cup lard
    1 cup crunchy peanut butter
    2 cups rolled oats
    2 cups cornmeal
    1 cup flour
    1 cup sugar

Melt the lard and peanut butter in a saucepan at low heat. Add the remaining ingredients and stir well to form a thick mixture. Spoon or pour the mixture into butter tubs (or into cookie cutters on a wax-paper–lined tray) to a depth of about 1 inch.

Poke a hole in the suet cake near the top, making it large enough to thread ribbon or string through. Freeze the suet for at least 1 week. Remove the suet cakes from the tubs or cookie cutters, peeling off any waxed paper. Thread ribbon or string through each suet cake and hang outside for the birds to enjoy.

Peanut-Coated Treats

    Creamy peanut butter
    Stale whole-grain bread, day-old doughnuts, or dense muffins
    Finely chopped peanuts or raisins

Working on one side of the bread, doughnut, or muffin at a time, spread a thin layer of peanut butter over it as you would icing. Press the peanut-buttered edge against the chopped peanuts, raisins, or cherries. Repeat for the remaining surfaces. Offer these bird-friendly treats to the birds on a platform feeder or hang them with a ribbon from a tree branch.

Top 10 Beneficial Bugs You'll Want in Your Garden Next Year

These suggestions are sure to bring lots of beneficial insects and pollinators into your organic garden.

1. **Aphid midge:** The larvae of this tiny, long-legged fly feed on more than 60 species of aphids by paralyzing their prey with toxic saliva. Pollen plants will bring aphid midges to your garden.

2. **Braconid wasps:** The adult female of the species injects its eggs into host insects. The larvae then feed inside their hosts, which include moth and beetle larvae and aphids. The host dies once the larvae have completed development. Grow nectar plants with small flowers, such as dill, parsley, wild carrot, and yarrow, to bring them to your garden.

3. **Damsel bugs:** Damsel bugs feed on aphids, small caterpillars, leaf-hoppers, thrips, and other pesky pests. Collect damsel bugs from alfalfa fields, using a sweep net, and then release them around your site.

4. **Ground beetles:** The nocturnal ground beetle is a voracious predator of slugs, snails, cutworms, cabbage maggots, and other pests that live in your garden's soil. One beetle larva can eat more than 50 caterpillars! Plant perennials among garden plants for stable habitats, or white clover as a groundcover in orchards.

5. **Lacewings:** Both adult lacewings and their larvae eat aphids, caterpillars, mealybugs, scales, thrips, and whiteflies. Angelica, coreopsis, cosmos, and sweet alyssum will bring lacewings to your garden.

6. **Lady beetles:** Adult lady beetles eat aphids, mites and mealybugs, and their hungry larvae do even more damage to garden pests. Plant angelica, coreopsis, dill, fennel, and yarrow to attract them.

7. **Minute pirate bugs:** The quick-moving, black-and-white patterned minute pirate bugs will attack almost any insect. Goldenrods, daisies, alfalfa, and yarrow will attract these bugs.

8. **Soldier beetles:** The soldier beetle feeds on aphids, caterpillars, and other insects, including harmless and beneficial species. Attract this flying insect by planting catnip, goldenrod, and hydrangea.

9. **Spined soldier bug:** The spined soldier bug's pointed "shoulders" distinguish it from the peskier stink bug. Plant permanent beds of perennials to provide shelter for this predator of hairless aterpillars and beetle larvae.

10. **Tachinid flies:** Tachinid fly larvae burrow their way into many caterpillars, destroying these garden pests from the inside. Plant dill, parsley, sweet clover, and other herbs to attract adult flies.

GARDENER TO GARDENER

Definitely a Deer Deterent

I've found that bath soap makes a very effective deer repellent—it has kept them away from my fruit trees, garden phlox, roses, and columbine. I use duct tape to attach the soap—one bar for each small tree and several bars for larger trees—at a level of 3 to 6 feet above ground. Replace the bars as the weather dissolves them.

> Cheri Wujek
> Galesburg, Illinois

Stinky Socks Sock It to 'Em

My garden is in a wooded location, so I've had problems with deer and raccoons eating everything I plant. On the advice of a friend, I soaked some old socks in a mixture of six eggs (to repel the deer), some black pepper (to repel the raccoons), and enough water to make a milky-looking liquid. I then tied the socks to a string that went all around the perimeter of the garden. One "soaking of the socks" has kept the deer and raccoons away from the garden for the entire growing season.

> Mary Benjamin
> Willseyville, New York

Frighten Garden Foes

Last season, I inadvertently found a way to stymie the deer in my garden. I emptied a bag of leaves that I'd stored over the winter, then placed the smelly, empty bags inside out and upside down on tall stakes to air out and dry. I never got around to putting the bags away, and the wind rustled them day and night. Sometimes the sound startled me as I worked outdoors, and it must have startled the deer too because they stayed away from my garden all summer! In October, when I removed the tattered bags, the deer returned almost immediately.

> Gladys Uitvlugt
> McBain, Michigan

Tasty Alternatives

To prevent deer, rabbits, groundhogs, and other critters from eating my roses, vegetables, and shrubs, I leave a section of the property in tall, hardy fescue grass. The deer love to graze and forage in this area (on the side of the house opposite my flowers), and they leave all of my treasured plantings alone.

> Matthew Thomas Jr.
> Lynchburg, Virginia

Hair Today, Gone Tomorrow

The best thing I've found for keeping deer—and other "human-wary" pests—out of the garden is human hair. Make small bags out of netting or old pantyhose, fill them with fresh hair, then hang the bags on a fence post or in a tree, or put them right on the ground. If you put out enough of these little packets, deer will turn away. Occasionally renew the bags with fresh hair. If you don't have enough of your own hair on hand, ask a local barber for some clippings.

> Kathleen Wright
> Glendale, Oregon

GARDENER TO GARDENER

Gone with the Wind

To keep deer out of our pea patch, we hammered in a few stakes, ran string around our peas, and tied strips of plastic to the string. The fluttering of the plastic strips in the wind scared the deer so much that they never bothered our peas.

Mrs. Howard D. Dalton
Gretna, Virginia

Good Choice for Pollinators

If you're interested in attracting pollinators to your garden, try planting 'African Blue' basil. It worked like magic to attract honeybees, mason bees, and other "good guys" to my garden last summer.

Gary Rose
Los Gatos, California

Moles and Gophers

To keep gophers from eating my prize bulbs, I plant the bulbs in "baskets" made of ½-inch wire mesh. Be sure to place the baskets deep enough in the planting hole so that they don't restrict root growth.

Sue Dawson
Muscoy, California

Decorate to Deter

For the past 3 years, we have been scaring deer away with colorful plastic pennants, such as those used by stores, used-car lots, and gas stations. We string the pennants between posts around the garden. Now we're able to harvest all our vegetables before the deer get to them.

Gladys Shippman
Redwood Falls, Minnesota

Sifting Out the Squirrels

I'd been battling squirrels for 11 years and couldn't find anything to keep them from ruining or stealing my tomatoes and peppers. Then, a local garden center employee recommended bloodmeal, which is easy to apply with a small hand sifter (a broadcast seed will work, too). I dust it on fruits and leaves and around the outside of the garden. Of course, you have to resprinkle after a good rain.

Dorothy R. Mann
Dayton, Ohio

Hare-Raising Tale

To prevent rabbits from feeding on tender bean and pea seedlings, I place dog hair next to the plants, anchoring small amounts of the hair with stones or sticks. The hair will work for several weeks—even after a rain.

Dave Asendorf
Armacost, Maryland

Dogged Defenders

In my neighborhood we have an abundance of squirrels that love to eat ripening tomatoes. But they pass right by my yard without a glance, thanks to my two large dogs, which patrol the outside of the garden. I surrounded the garden with a small fence to keep out the dogs, and it has worked great.

Joelle Larkworthy
Kansas City, Missouri

GARDENER TO GARDENER

Peppers Ward Off Moles

I've found that hot peppers work great for repelling moles. I stuffed several dozen jalapeno peppers into mole burrows at intervals of 2 to 3 feet, then packed the earth back down firmly, and didn't see any further damage in the area for the rest of the season. Now I keep a bag of jalapeno peppers in the freezer to ensure that visiting moles do not become permanent residents.

Debora Granneman
Troy, Michigan

It's Electrifying

I've tried many things to keep the raccoons and other wild animals out of my garden, but none has worked as well as the electric fence that I put up about 10 years ago. I strung two wires—one 6 inches from the ground, the other 6 inches above that—on wooden stakes spaced about 20 feet apart.

Joseph P. Smith
Enfield, Connecticut

Garlic Mulch

I've discovered that raccoons dislike garlic. Every year we grow hundreds of flower seedlings and set them into the ground after the danger of frost has passed. But when we checked on our transplants in the morning, we often found the young plants uprooted, flung about, mangled, and sometimes buried.

One day I snipped some green tops from our garlic plants and sprinkled them on and around my transplants. The next morning, only one flower had been uprooted. We replanted it, and mulched around it with more garlic—an extra-large handful. That, my friends, was that—no more uprooted plants. When I suffered a temporary lapse of diligence and forgot to put garlic clippings on a row of mixed flowers, the raccoons returned that same night. That was the last time I forgot to mulch my transplants with garlic tops.

Debi Larson
Mountain View, Missouri

Lure House Sparrows Away

I try hard to attract many different species of birds to my garden, and it used to be frustrating to watch house sparrows outmuscle the other birds at the feeder. Now I lure the pesky sparrows to a different feeding station—so that the chickadees, woodpeckers, tufted titmice, nuthatches, and goldfinches have a place of their own to chow down:

First, I scatter some bird feed on the ground about 10 feet away from the regular feeding area. Then—because house sparrows are very wary of new objects—I fasten strips of colorful cloth, paper, foil, and old bread wrappers onto my regular feeding station.

If you're using a platform feeder, just set an old can, colorful tumbler, or even an old boot on the platform to scare away the sparrows. These scare tactics don't seem to bother the other, more desirable species as long as the objects don't get in the way of the openings on the feeder. Meanwhile, the sparrows settle for their food across the yard, where the seed is scattered on the ground.

Anna Mae Plank
Arthur, Illinois

GARDENER TO GARDENER

Fairies and Felines

I have managed to keep cats away from my bird feeder by planting three rose bushes ('The Fairy') beneath it. 'The Fairy' is very hardy and easy to grow, and it spreads without growing too tall so refilling the feeder isn't a problem. I also planted purple clematis in the middle. It climbs the pole and looks pretty in combination with the pink rose clusters. The clematis does especially well in this spot because this variety likes to have "its feet in the shade and its head in the sun." ('The Fairy' rose is widely available at garden centers and from mail-order suppliers of roses.)

Teresa Kucera
Epping, New Hampshire

Dandelions for the Birds

Dandelions are one of the more appreciated weeds in our chemical-free lawn. During the growing season, I send my youngsters out in the afternoon to collect the pretty yellow seedheads and put them into paper bags. I stash the bags in a dark cupboard until after Thanksgiving. Then, I take them out and fill our birdfeeders with the stored seedheads. Birds love the dandelions. At the same time, the children learn how to harvest a crop, and they discover that weeds are not useless.

Donna Thompson
Pe Ell, Washington

Scat, Deer!

Got problems with deer eating your prized ornamentals? Well, here's a way to send them packing. Go to the nearest zoo, visit the large cat exhibit (take a large trash bag with you), and ask the zookeeper for a bagful of mountain lion scat (cat poop). Actually, scat from any large cat will do, but a mountain lion's seems to work the best, and the zoos are usually happy to get rid of it.

Now take the bag home and spread the contents around the perimeter of your flower garden. I haven't seen a deer go near the stuff! (Note: For safety, avoid using fresh manure of any type around food plants.)

James Glennon
Malibu, California

Bee Sweet to Pollinators

Last April, when I didn't spot a single bee around my blossoming apple trees, I decided to entice them to come to "work" by offering one of their favorite foods—honey!

I put ½ cup of honey into a pitcher, filled the pitcher the rest of the way with warm water, and stirred the mixture well. At about 2 PM (when the wind usually picks up), I went to my orchard and tossed the mixture onto my favorite apple tree. I then went home and waited for the wind to carry my invitation to the bees. A short time later, I returned to inspect the blooms. Sure enough, my plan had worked! There were four blue orchard bees, three honeybees, and two black flies, all busy pollinating the blossoms.

William F. Holden
West Linn, Oregon

Berry Nice Netting

In my small garden plot, I use inexpensive nylon netting to keep birds away from strawberries (and other animals out of green leafy vegetables). You can get the netting in a fabric store. Simply drape it over plants, like you would do with a row cover, and anchor the edges with soil.

Mary Lee DePew
Adrian, Michigan

GARDENER TO GARDENER

Vinegar Deters Spraying Cats

I live in a city co-op with a small front yard and an even smaller patio. The cats in the neighborhood used to spray the adjoining wall and stink up my yard to mark "their" territory—until I discovered a way to mark it myself.

I put some white vinegar in a small squirt bottle and applied it to the wall for three evenings in a row. The cats stayed away and the odor quickly disappeared.

A friend whose cats were clawing her furniture tried this too. She sprayed just enough vinegar into the air for the cats to get a good whiff. Now all she has to do is head in the cats' direction with the spray bottle and they head elsewhere in a hurry.

Victoria Price
Silver Spring, Maryland

Make a Better Mousetrap

Mice were the worst problem in my greenhouse. They'd dig in the seed flats and graze off the tops of my tiny seedlings. But I finally managed to halt their efforts with a no-fail trap made by a friend.

To make the trap you need a 5-gallon bucket, an empty soda can, one wire coat hanger, and some peanut butter for bait. Begin by clipping the hook off the hanger and straightening out the wire. Next, use a lighter or propane torch to heat one end of the wire until it's hot enough to melt two small holes into the sides of the bucket—one on each side about 2 inches below the rim. Use a pointy object to make two holes near the top of the soda can; space them directly opposite each other.

Now run the wire through one of the holes in the bucket, through both holes in the can, and out the hole in the other side of the bucket. Bend the ends of the wire into an L shape to keep them from slipping out of the bucket. Squeeze the can to make four dents equidistant from each other, then put some peanut butter into each of the dents.

Position the trap in a place where the little bandits will be able to reach the rim (I set mine next to a wooden step ladder; they scurry up the wood and do a trapeze act right into the bucket.) The mice will fall in, but will be unable to climb back out. What you do with your catch is your secret—just don't release them near my house!

Juanita Clark
Olympia, Washington

Gardener's Glossary

Acidic. Soil pH that's less than 7.0 (neutral); acidic soils tend to be deficient in phosphorus and sometimes contain excess manganese and aluminum.

Alkaline. Soil pH that's above 7.0 (neutral); alkaline soils tend to lack manganese and boron.

Annual. A plant that flowers, bears seed, and dies within one growing season.

***Bacillus thuringiensis* (BT).** Naturally occurring pathogen that's toxic to insect larvae but not harmful to other organisms or humans; various strains are sold commercially.

Backfill. To fill in a planting hole around a plant's roots with soil.

Bareroot. Plants (usually woody ornamentals, such as roses, shrubs, or young trees) sold without soil on their roots.

Beneficials. Insects considered helpful in the garden because they prey on pest insects. Examples: lady beetles, ground beetles, tachinid flies.

Blanch. 1. To mound soil around stems or leaves of growing plants, especially leeks and asparagus, so that they become white and tender. **2.** In food preservation, the process of boiling vegetables for a specific time period in order to slow deterioration during storage.

Bolt. To produce flowers and seed prematurely (often due to hot weather).

Broadcast. To scatter seed by hand in a random pattern.

Cell pack. Thin plastic or peat containers that have several compartments; used for starting individual seedlings.

Clay. Soil type that absorbs and drains water very slowly, due to lack of air space between particles.

Coldframe. Low, enclosed structure for protecting plants from cold; clear cover lets in sunlight.

Compost. A humus-rich, organic material formed by the decomposition of leaves, grass clippings, and other organic materials. Used to improve soil.

Cover crop. Plant grown and then turned under to improve soil texture and fertility. Examples: clover, rye, buckwheat, vetch.

Crown. The point where a plant's roots and stem meet, usually at soil level.

Cultivar. A cultivated variety of a plant, usually selected for a special trait, such as compact growth or disease resistance.

Cutting (hardwood). Mature wood (deciduous or evergreen) taken at the end of the growing season or during dormancy in order to start new plants.

Deadhead. To remove faded flowers.

Determinate tomatoes. Tomatoes that have vines that grow to a given height and then stop. Most of the fruit is produced and ripens at one time.

Diatomaceous earth. The fossilized remains of ancient marine organisms, sold as a control for soft-bodied insect pests, such as aphids and slugs.

Direct seed. To sow seeds outdoors in garden soil (as opposed to starting seed indoors for transplanting at a later date).

Double dig. To work soil to a depth that is twice the usual by digging a trench, loosening the soil at the bottom of the trench, then returning the top layer of soil to the trench. This produces a raised bed that contains a deep layer of very loose, fluffy soil.

Drip irrigation. A type of irrigation in which water seeps slowly into the soil via hoses or pipe systems. There is less evaporation and more control than with overhead watering.

Drip line. Imaginary line on the soil marking the outside circumference of a tree's canopy.

Flat. Shallow tray for starting seeds indoors; does not have individual compartments.

Force. To bring dormant plants (especially bulbs and cut branches) into bloom by manipulating temperature.

Frass. Insect excrement; can be used to help identify garden pests.

Friable. Having a crumbly texture. Desirable for soil.

Green manure. A fast-maturing leafy crop—such as buckwheat, rye, or clover—that adds organic matter to the soil when it's turned under.

Harden off. Gradually expose a plant to outdoor conditions before transplanting it to the garden.

Hardiness. Ability to survive the winter without protection from the cold.

Humus. The complex, organic residue of decayed plant matter in soil.

Hybrid. The offspring of genetically different plants.

Indeterminate tomatoes. Tomato varieties that produce vines that continue to grow for the life of the plant.

Interplanting. Combining plants that have different bloom times or growth habits so that bloom lasts longer and chance of disease is lower.

Leaf mold. Decomposed leaves. An excellent winter mulch, attractive to earthworms.

Loam. An ideal garden soil; contains plenty of organic matter and a balanced mix of small and large particles.

Medium. Soil mix or potting mixture.

Microclimate. Local conditions of shade, exposure, wind, drainage, and other factors that affect plant growth at a given site.

Mulch. Material layered over the soil's surface to hold in moisture, suppress weeds, and (if organic) improve soil.

Open-pollinated. Varieties whose seeds come from plants pollinated naturally.

Organic matter. Materials derived from plants and animals, such as leaves, grass clippings, and manure.

Overwinter. To keep a plant living through the winter so that it can continue growing the next year; may require the use of coldframes, row covers, or hoop houses.

Pathogen. Disease-causing microorganism.

Peat pot. Commercially available seedling container made from compressed peat moss. Gradually breaks down, so both pot and seedling can be planted without disturbing roots.

Perennial. Any plant that lives for at least three seasons; can be woody (trees and shrubs) or herbaceous (those that die back to the ground in winter).

Perlite. Lightweight expanded minerals added to potting mixes to improve aeration.

pH. Measure of soil acidity or alkalinity; 7.0 is neutral.

Pinching. Periodically removing the newest growth of a leafy plant to encourage a more "bushy" form.

Pot-bound. A containerized plant whose roots have completely filled the potting soil inside the container, making it difficult to remove the plant from the container. Pot-bound plants may have stunted growth.

Pot up. To plant in a container or to move a containerized plant into a larger container.

Row cover. Translucent polyester fabric placed over garden plants or beds to protect them from pests. Heavier types also used to protect plants from cold. Lets in water and light.

Side-dress. To spread a layer of compost or other fertilizer on the soil surface, next to growing plants.

Succession crop. To follow early-maturing crops with other plantings so that beds continue to produce throughout the growing season.

Thin. To remove excess seedlings or fruits to ensure best spacing for plant health, yield, and size.

Tilth. Soil texture and workability.

Top-dress. To apply a layer of fertilizer to the soil surface (not working it in).

Umbel-shaped. Plants with flower heads shaped like an umbrella, including dill, fennel, anise, yarrow, and coriander. Attractive to beneficial insects.

Underplant. To plant short plants, such as ground covers, below taller plants, such as shrubs.

Vermiculite. Lightweight mineral added to soil mix to improve aeration.

Wallo'Water. A product sold to protect plants from cold; water-filled plastic walls retain heat around plants.

Zones. Geographic regions; hardiness zones marked by a range of lowest temperatures in an average winter for a given area. Example: 0° to −10°F.

Resources

Use the listings below to find companies that sell seeds, plants, and organic gardening supplies, as well as organizations that offer expertise in birdwatching. When you contact associations or specialty nurseries by mail, please enclose a self-addressed, stamped envelope with your inquiry.

Vegetable and Flower Seeds

Bountiful Gardens
18001 Shafer Ranch Road
Willits, CA 95490
Phone: (707) 459-6410
Fax: (707) 459-1925
Web site: www.bountifulgardens.org

W. Atlee Burpee
300 Park Avenue
Warminster, PA 18974
Phone: (800) 888-1447
Fax: (215) 674-4170
Web site: www.burpee.com

The Cook's Garden
PO Box 535
Londonderry, VT 05148
Phone: (800) 457-9703
Fax: (800) 457-9705
Web site: www.cooksgarden.com

Fedco Seeds
PO Box 250
Waterville, ME 04903
Phone: (207) 873-7333
Fax: (207) 872-8317

Garden City Seeds
778 Highway 93 North, #3
Hamilton, MT 59840
Phone: (406) 961-4837
Fax: (406) 961-4877
Web site: www.gardencityseeds.com

Johnny's Selected Seeds
1 Foss Hill Road
RR 1, Box 2580
Albion, ME 04910
Phone: (207) 437-4301
Fax: (207) 437-2165
Web site: www.johnnyseeds.com

Nichols Garden Nursery
1190 Old Salem Road NE
Albany, OR 97321
Phone: (541) 928-9280
Fax: (800) 231-5306
Web site: www.nicholsgardennursery.com

Park Seed Co.
1 Parkton Avenue
Greenwood SC 29647
Phone: (800) 845-3369
Fax: (864) 941-4206
Web site: www.parkseed.com

Pinetree Garden Seeds
PO Box 300
New Gloucester, ME 04260
Phone: (207) 926-3400
Fax: (888) 527-3337
Web site: www.superseeds.com

Seeds of Change
PO Box 15700
Santa Fe, NM 87506
Phone: (888) 762-7333
Web site: www.seedsofchange.com

Southern Exposure Seed Exchange
PO Box 460
Mineral, VA 23117
Phone: (540) 894-9480
Fax: (804) 894-9481
Web site: www.southernexposure.com

Stokes Seed Inc.
Box 548
Buffalo, NY 14240
Phone: (716) 695-6980
Fax: (888) 834-3334
Web site: www.stokeseeds.com .

Territorial Seed Company
PO Box 158
Cottage Grove, OR 97424
Phone: (541) 942-9547
Fax: (888) 657-3131
Web site: www.territorial-seed.com

Tomato Growers Supply Co.
PO Box 2237
Fort Myers, FL 33902
Phone: (888) 478-7333
Fax: (888) 768-3476
Web site: www.tomatogrowers.com

Fruit

Bear Creek Nursery
PO Box 411
Bear Creek Road
Northport, WA 99157
Phone: (509) 732-6219
Fax: (509) 732-4417

Johnson Nursery, Inc.
5273 Highway 52 East
Ellijay, GA 30540
Phone: (888) 276-3187
Fax: (706) 276-3186
Web site: www.johnsonnursery.com

Lewis Nursery & Farms, Inc.
3500 NC Hwy 133 West
Rocky Point, NC 28457
Phone: (910) 675-2394
Fax: (910) 602-3106

Raintree Nursery
391 Butts Road
Morton, WA 98356
Phone: (360) 496-6400
Fax: (888) 770-8358
Web site: www.raintreenursery.com

Street Lawrence Nurseries
325 SH 345
Potsdam, NY 13676
Phone: (315) 265-6739
Web site: www.sln.Potsdam.ny.us

Strawberry Tyme Farms
RR 2 #1250, Street Johns Road
Simcoe, ON
N3Y 4K1 Canada
Phone: (519) 426-3009
Fax: (519) 426-2573
Web site: www.strawberrytyme.com

Perennials

André Viette Farm & Nursery
PO Box 1109
Fishersville, VA 22939
Phone: (540) 943-2315
Fax: (540) 943-0782
Web site: www.viette.com

Bluestone Perennials
7211 Middle Ridge Road
Madison, OH 44057
Phone: (800) 852-5243
Fax: (440) 428-7535
Web site: www.bluestoneperennials.com

Busse Gardens
17160 245th Avenue
Big Lake, MN 55309
Phone: (800) 544-3192
Fax: (612) 263-1473
Web site: www.bussegardens.com

Canyon Creek Nursery
3527 Dry Creek Road
Oroville, CA 95965
Phone: (530) 533-2166
Web site: www.canyoncreeknursery.com

Carroll Gardens
444 E. Main Street
Westminster, MD 21157
Phone: (800) 638-6334
Fax: (410) 857-4112
Web site: www.carrollgardens.com

Forestfarm
990 Tetherow Road
Williams, OR 97544
Phone: (541) 846-7269
Fax: (541) 846-6963
Web site: www.forestfarm.com

Gardens North
5984 Third Line Road North
North Gower, ON
K0A 2T0 Canada
Phone: (613) 489-0065
Fax: (613) 489-1208
Web site: www.gardensnorth.com

Heronswood Nursery Ltd.
7530 NE 288th Street
Kingston, WA 98346
Phone: (360) 297-4172
Fax: (360) 297-8321
Web site: www.heronswood.com

Kurt Bluemel, Inc.
2740 Greene Lane
Baldwin, MD 21013
Phone: (800) 248-7584
Fax: (410) 557-9785
Web site: www.bluemel.com

Niche Gardens
1111 Dawson Road
Chapel Hill, NC 27516
Phone: (919) 967-0078
Fax: (919) 967-4026
Web site: www.nichegdn.com

The Perennial Gardens
13139 224th Street
Maple Ridge, BC
V4R 2P6 Canada
Phone: (604) 467-4218
Fax: (604) 467-3181
Web site: www.perennialgardener.com

Plant Delights Nursery
9241 Sauls Road
Raleigh, NC 27603
Phone: (919) 772-4794
Fax: (919) 662-0370
Web site: www.plantdel.com

Plants of the Southwest
Agua Fria, Route 6, Box 11A
Santa Fe, NM 87501
Phone: (800) 788-7333
Fax: (505) 438-8800
Web site: www.plantsofthesouthwest.com

Shady Oaks Nursery
PO Box 708
Waseca, MN 56093
Phone: (800) 504-8006
Fax: (888) 735-4531
Web site: www.shadyoaks.com

Siskiyou Rare Plant Nursery
2825 Cummings Road
Medford, OR 97501
Phone: (541) 772-6846
Fax: (541) 772-4917
Web site: www.wave.net/upg/srpn

Wayide Gardens
1 Garden Lane
Hodges, SC 29695
Phone/Fax: (800) 845-1124
Web site: www.waysidegardens.com

White Flower Farm
PO Box 50
Litchfield, CT 06759-0050
Phone: (800) 503-9624
Fax: (860) 496-1418
Web site: www.whiteflowerfarm.com

Gardening Supplies/Soil Testing

Arbico
18701 North Lago Del Oro Pkwy
Tucson, AZ 85739
Phone: (800) 827-2847
Web site: www.goodearthmarketplace.com

Gardens Alive!
5100 Schenley Place
Lawrenceburg, IN 47025
Phone: (812) 537-8650
Fax: (812) 537-5108
Web site: www.gardensalive.com

Gardener's Supply Co.
128 Intervale Road
Burlington, VT 05401
Phone; (888) 833-1412
Fax: (800) 551-6712
Web site: www.gardeners.com

Green Spot Ltd.
93 Priest Road
Nottingham, NH 03290
Phone: (603) 942-8925
Fax: (603) 942-8932
Web site: www.greenmethods.com

Harmony Farm Supply
3244 Highway 116 North
Sebastopol, CA 95742
Phone: (707) 823-9125
Fax: (707) 823-1734
Web site: www.harmonyfarm.com

Peaceful Valley Farm Supply
PO Box 2209
Grass Valley, CA 95945
Phone: (530) 272-4769
Fax: (530) 272-4794
Web site: www.groworganic.com

Snow Pond Farm Supply
PO Box 70
Salem, MA 01970
Phone: (978) 745-0716
Fax: (978) 745-0905
Web site: www.snow-pond.com

Woods End Research Laboratory
PO Box 297
Mt. Vernon, ME 04352
Phone: (800) 451-0337
Fax: (207) 293-2488
Web site: www.solvita.com

Worm's Way
7850 N. Highway 37
Bloomington, IN 47404
Phone: (800) 274-9676
Web site: www.wormsway.net

Seed-Saving Organizations

Maine Seed-Saving Network
PO Box 126
Penobscot, ME 04476
Phone: (207) 326-0751

Seeds of Diversity Canada
PO Box 36, Station Q
Toronto, ON
M4T 2L7 Canada
Phone: (905) 623-0353
Web site: www.seeds.ca

Seed Savers Exchange
3076 N. Winn Road
Decorah, IA 52101
Phone: (319) 382-5990
Fax: (319) 382-5872
Web site: www.seedsavers.org

Books and Periodicals

Ashworth, Suzanne. *Seed to Seed: Seed Saving Techniques for the Home Gardener*. Decorah, IA: Seed Saver Publications, 1991.

Bradley, Fern Marshall, and Barbara W. Ellis, eds. *Rodale's All-New Encyclopedia of Organic Gardening*. Emmaus, PA: Rodale, 1992.

Coleman, Eliot. *Four-Season Harvest: Organic Vegetables from your Home Garden All Year Around*. White River Junction, VT: Chelsea Green, 1999.

DiSabato-Aust, Tracy. *The Well-Tended Perennial Garden*. Portland, OR: Timber Press, 1998.

Druse, Ken. *Making More Plants: The Science, Art, and Joy of Propagation*. New York: Clarkson Potter, 2000.

Editors, Rodale Organic Gardening Books. *Rodale Organic Gardening Basics: Roses*. Emmaus, PA: Rodale, 2000.

Gilkeson, Linda, Pam Peirce, and Miranda Smith. *Rodale's Pest & Disease Problem Solver: A Chemical Free Guide to Keeping Your Garden Healthy*. Emmaus, PA: Rodale, 1996.

Jeavons, John. *How to Grow More Vegetables Than You Ever Thought Possible on Less Land Than You Can Imagine: A Primer on the Life-Giving Biointensive Method of Organic Horticulture*. Berkeley, CA: Ten Speed Press, 1991.

Martin, Deborah, and Grace Gershuny, eds. *The Rodale Book of Composting*. Emmaus, PA: Rodale, 1992.

Moyer, Anne Halpin. *Foolproof Planting: How to Successfully Start and Propagate More Than 250 Vegetables, Flowers, Trees, and Shrubs*. Emmaus, PA: Rodale, 1990

Nick, Jean M. A., and Fern Marshall Bradley. *Growing Fruits & Vegetables Organically*. Emmaus, PA: Rodale, 1994.

Ondra, Nancy J. *Soil and Composting*. (Taylor's Weekend Gardening Guides.) Boston: Houghton Mifflin, 1998.

OG magazine, Rodale Inc., Emmaus, PA 18098.

Phillips, Ellen, and C. Colston Burrell. *Rodale's Illustrated Encyclopedia of Perennials*. Emmaus, PA: Rodale, 1993.

Powell, Eileen. *From Seed to Bloom*. Pownal, VT: Storey, 1995.

Sombke, Laurence. *Beautiful Easy Flower Gardens*. Emmaus, PA: Rodale, 1995.

Index

USDA Plant Hardiness Zone Map

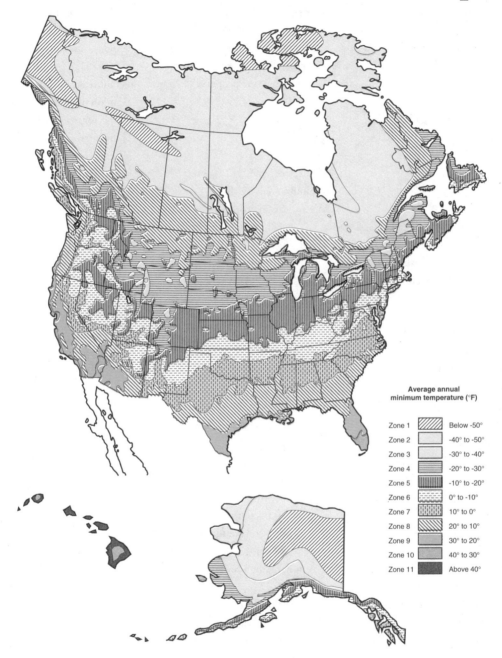

Average annual minimum temperature (°F)

Zone 1	Below -50°
Zone 2	-40° to -50°
Zone 3	-30° to -40°
Zone 4	-20° to -30°
Zone 5	-10° to -20°
Zone 6	0° to -10°
Zone 7	10° to 0°
Zone 8	20° to 10°
Zone 9	30° to 20°
Zone 10	40° to 30°
Zone 11	Above 40°

This map was revised in 1990 and is recognized as the best indicator of minimum temperatures available. Look at the map to find your area, then match its color to the key. When you've found your color, the key will tell you what hardiness zone you live in. Remember that the map is a general guide; your particular conditions may vary.